Screen People >

Megan Garber

SCREEN PEOPLE

How We Entertained Ourselves into a State of Emergency

WILDFIRE

Copyright © Megan Garber 2026

The right of Megan Garber to be identified as the Author of the Work has been asserted by her in accordance with the Copyright, Designs and Patents Act 1988.

Published in agreement with HarperOne

An imprint of HarperCollins Publishers

Designed by Yvonne Chan

User icon illustration © Maksim/stock.adobe.com

Title page art by Ben Wiseman

Mobile status bar icons © Cezar_911/stock.adobe.com

First published in Hardback in Great Britain in 2026 by Wildfire
An imprint of Headline Publishing Group Limited

1

Apart from any use permitted under UK copyright law, this publication may only be reproduced, stored, or transmitted, in any form, or by any means, with prior permission in writing of the publishers or, in the case of reprographic production, in accordance with the terms of licences issued by the Copyright Licensing Agency.

Cataloguing in Publication Data is available from the British Library

Hardback ISBN 978 1 0354 3045 1
Trade Paperback ISBN 978 1 0354 3046 8

Offset in 13.7/19.4pt Adobe Caslon Pro by Six Red Marbles UK, Thetford, Norfolk

Printed and bound in Great Britain by Clays Ltd, Elcograf S.p.A.

Headline's policy is to use papers that are natural, renewable and recyclable products and made from wood grown in well-managed forests and other controlled sources. The logging and manufacturing processes are expected to conform to the environmental regulations of the country of origin.

Headline Publishing Group Limited
An Hachette UK Company
Carmelite House
50 Victoria Embankment
London EC4Y 0DZ

The authorised representative in the EEA is Hachette Ireland,
8 Castlecourt Centre, Dublin 15, D15 XTP3, Ireland (email: info@hbgi.ie)

www.headline.co.uk
www.hachette.co.uk

For Corey Garber, a boundless person

Contents

1
Introduction

31
1: The Screens

56
2: The Stars

90
3: The Sets

123
4: The Scripts

141
5: The Producers

176
6: The Extras

192
7: The Fans

206
8: The Haters

241
9: The Twists

274
10: The End

285
Acknowledgments

Screen People >

Introduction

February 26, 2015, was an exceptionally internet-y day on the internet. In a town near Phoenix, late morning local time, two llamas serving as therapy animals at a retirement community broke away from their handlers. Soon, they were galloping through the streets and yards of suburban Sun City, Arizona, their necks outstretched, their fluff streaking in the breeze. They were trailed by seniors in golf carts and, eventually, by horseback riders who tossed lassos toward the wayward camelids.

The chase would soon be televised. Local TV stations, knowing that "llamas on the lam" was a human-interest story at heart, sent helicopters to capture the scene. The images spread, first to the stations' national affiliates and then to the network at large. The llamas, as they galumphed through Sun City, sped across the country, as well, through grainy videos and sassy memes. They were Thelma and Louise. They were Bonnie and Clyde. They were photogenic in an especially web-friendly way: They seemed, as *The New York Times* would later note, "jaunty and defiant."

#Llamadrama, as viral stories went, was about as wholesome as it gets. Here was a low-speed chase with refreshingly low stakes, offering distraction and delight on an otherwise standard-issue Thursday. The animals—made way for by cars and pedestrians alike—were never really in danger: They were soon returned, safely, to their handlers, their allotted fifteen minutes having extended to several hours of fame. Their human pursuers had grinned gamely for the cameras, seeming aware that the golf carts and lassos had transformed the llamas' buddy comedy into a farce.

And then came Act II. Later that Thursday, the *BuzzFeed* writer Cates Holderness republished a picture she had come across on Tumblr: a dress, horizontally striped and topped with a matching bolero, hanging against a window. The post had a simple headline—"What Colors Are This Dress?"—and a brief explanation: "There's a lot of debate about the color of the dress." It concluded with a poll inviting readers to weigh in. Was the garment in question black and blue . . . or white and gold?

The dress fast acquired the abbreviated name recognition typically reserved for megastars: It became, simply, The Dress. The garment *broke the internet*, as people said back then, because it also broke people's minds. Some saw one set of colors. Some saw another. There was no nuance to be found in this, no gray area between the blue-and-black and the white-and-gold. Teams formed and clashed, each side (1) absolutely sure that it saw the dress as it really was, and (2) baffled by the other side's delusion. Houses divided, melodramatically, their members speaking of divorces and disownments and feuds that would extend, Hatfield-and-McCoy-style, to future generations. Celebrities weighed in. So did politicians. So did God (or, at any rate, the widely followed Twitter account @TheTweetOfGod that spoke to humanity in max-140-character bursts): "The color of a dress? Really? That's what you're asking Me?" it thundered.

But people were not strictly questioning the color of the dress. They were questioning the nature of reality. "WHAT IN THE WORLD IT IS WHITE AND GOLD," a group-chat exchange might start. To which: "wait are you serious ITS BLUE AND BLACK." Everyone was an eye witness. Everyone's testimony was suspect. The Dress, a mystery wrapped in an enigma wrapped in a sassy bolero, had brought division but no explanation. The conversation about it expanded accordingly—toward cosmic musings (singularities, simulations, red pills, blue), conspiracy theories (*it's a hoax! no, it's an Illuminati mind trick!*), and various flavors of all-caps existential panic: "UGGGHHHH WHAT IS HAPPENING? IS THIS A JOKE??????"

The Dress, however, was not a joke. It was simply a trick of the eye—and a product of the weird things that can happen when the human body interacts with the screen. The light emitted by phones and computers is distinct from the light of the sun and the light of the bulb. It travels at different wavelengths, and hits the eye differently as a result. When exposed to screen light for prolonged periods, the eyes' receptors adjust to the new conditions—as they might when, say, you move from a dim room to a brightly lit one. The distinction between #teamblueandblack and #teamwhiteandgold, whatever else it might reveal—hero or villain, Gryffindor or Slytherin, the right side of history or the wrong—was a matter of screen time and random circumstance. The color you saw, for the most part, was a reflection of how long you'd been interacting with screens before your eyes met The Dress. If your eyes had already adjusted to the "environment" of artificial light, you would likely see it as white and gold. If you hadn't, you would likely see it as it was in life: black and blue.

The Dress, having ruptured reality so arbitrarily, might have lived on as a cautionary tale—an ever-timely reminder of the constraints of our vision and the limits of our certainties. Instead, for the most part, it lives on as lore. News sites mark its anniversaries. Pub-trivia writers weave it into their quizzes. They sometimes recall the day's fractures; mostly, though, they remember its fun. That's how I remember it, too. The partisanship The Dress provoked was playful, in the end, its vitriol a good-natured performance. People could hurl their all-caps invectives knowing that the digital shouts would not be mistaken for real ones. They could spin wild conspiracy theories knowing that the people on the receiving end would sense, if not see, their winks. They—we—could treat partisanship as a punch line, trusting that we were all in on the same joke.

The Dress arrived during a pivotal moment for the internet. By the winter of 2015, the World Wide Web, the system of hyperlinked pages that made the internet accessible to the public, had reached a generational milestone: It had turned twenty-five years old. Facebook and similar social-media platforms, then roughly a decade old, were completing their transition from novelties to banalities to anxieties. "Move fast and break things," Facebook's internal motto, by the winter of 2015, had become a matter of public knowledge. Learning of it, people wondered whether they might be among the broken.

The Dress, against that backdrop, seemed to make optimism an option again. All that weirdness, all that warmth—it had to mean something. But The Dress was that rare cultural event that, having attracted a mass audience, had no real reboots or sequels. Instead, it seemed to herald an era of ever more fractures and ever less fun. People grew angrier, louder, lonelier, wearier. When they screamed through their keyboards, they meant it. When they appointed new enemies, they held the grudge. They were ever more suspicious. They were ever more certain. The Dress came to seem like a relic—a reminder of a time that was and would never be again. February 26, 2015, the explainer site *Vox* would write nearly a decade later, was "the last good day on the internet."

What changed, precisely, between the day of The Dress and this one? Why do so many people today feel homesick for a place they never left?

The answer I've come to is related to the one that explained why, on that hectic Thursday in 2015, people across the internet looked at one dress and saw two: Screens change things. They do it obviously. They do it subtly. They do it inevitably. Screens make the far things seem near. They make the near things seem far. They transform "UGGGHHHH WHAT IS HAPPENING??????" from a daylong joke into an endless anxiety. They convey the world. They capture it. They distort it. They destabilize it.

The premise of this book is that they do the same thing to people.

Mediums and Messages

The television, in the twentieth century, rearranged the American living room. In the pre-TV era, family spaces were typically furnished to accommodate conversation: couches and chairs next to each other and across from each other, allowing their occupants to chat and play and, in general, serve as each other's entertainment. In the 1950s and '60s, though, another pattern became the norm. Seats faced the new TV sets, fanning out around them with centrifugal precision, angled to give everyone the best possible view of the screen. The change was quick, decisive, and enduring: People used to look at one another. Now, they watched the TV.

Decades after those first TV sets made their way into American homes, people have found ever more ways to orient themselves toward the screens. TVs preside over airport lounges, hotel lobbies, taxicabs, bedrooms, bathrooms, kitchens, dinner tables. Screens now hang on walls like pieces of art. The arrival of laptops and smartphones has done little to compromise their ubiquity: In response to all the compact screens, the televisions have merely gotten bigger.

"Tele" means "distance." Its combination with "vision," in the early twentieth century, conveyed the wonder of an invention that made distant images newly portable. The wonder of today's screens, though, is very often their intimacy. They hear our breaths and hold our heartbeats. They come to life at the warmth of our touch. They are things we watch and wear, yes, but they are also extensions of our bodies and minds. They are portals. They are appendages. They are proof, that is to say, that Marshall McLuhan was right.

The Canadian theorist's most famous aphorism, "The medium is the message," was a gauzy phrase with a direct point: The technologies people use to communicate are more than technologies. They are more than tools. They are more than straightforward conveyors of content. Mediums shape the content, too. Along the way, they shape the humans who use them. Mediums, for McLuhan, were "extensions of man." They were embodiments of the declaration made by his friend, the scholar John Culkin: "We shape our tools and thereafter our tools shape us."

McLuhan introduced his theories in the 1960s, when TV was finding hegemony and when print newspapers were giving circadian shape to Americans' lives. When he talked about mediums, information systems were thoroughly one-way in their workings. Newspapers and books offered text; the audiences read them. Radios offered audio; the audiences listened. The screens offered images; the audiences consumed them.

But the screens of the internet, however much they might resemble the screens of the television, work differently. They are two-way affairs. Their images might come to us as before from faraway broadcast booths, zapping with electric ease into view; just as often, though, they are born of other people. The screens are interactive. They are discursive. We no longer merely watch the screens; we also exist within them.

And that difference, in turn, affects everything else.

On our two-way screens, we become two-way people. We are humans, and we are images. We are *who*s and *what*s at once. We are, like The Dress, both real and mere tricks of the light. Converted into pieces of media, we might be edited. We might be cropped. We might be memed. We might be mocked. We might be rated. We might be shared. We might be deleted.

And we will have, in general, very little say in the matter. "Objectification," in the world at large, is widely recognized as a violation. In the world of screens, though, it is a fact of physics.

"Why are people so mean?" Donald Trump asked, in a 2022 social-media post. The answer is often attributed to politics: the partisanship, the widespread mistrust, the carnivalized cruelty that is neatly exemplified, as it happens, by the once and current president. But politics alone cannot account for our malignancies. Our fractures are matters of culture, too. They are consequences of lives lived on, and in, the screens. Screens make it ever easier to look at one another; they can also make it more difficult to see one another. They bring ever more banality, and ever more urgency, to the writer Aldous Huxley's observation about propaganda: Its core aim, he wrote, is to make one group of people forget that another group is human.

Forgetfulness becomes easier when the people themselves, as a matter of logistical necessity, have their three dimensions flattened into two. In the world beyond the screens—and, yet, ever more shaped by them—principles so basic that they are taught to kindergartners along with their ABCs—sharing; honesty; respect; the fact that, as the popular meme goes, *we live in a society*—have been so rapidly losing their purchase. "I Don't Know How to Explain to You That You Should Care About Other People," a 2017 *HuffPost* headline read. It went viral and for good reason: It was so eloquent, and so relatable, in its despair. It was heaving an exhausted sigh. It was sharing a primal scream.

This is one of the ironies—and in some sense one of the tragedies—of life in the internet age. For all the ways we now have to look at one another, we are becoming ever worse at really seeing one another. Interactive screens, in theory, should connect us. They should make us more legible to one another. But our new mediums muddy the message. Screens make us two things at once: subjects and objects, bodies and images, real and yet somehow not quite. Along the way, they make double vision a way of life.

The years that followed the winter of 2015 brought everyday forms of fracture. People sought attention and feared it. They compared it to a viral load. They argued in the comments sections. They argued everywhere else. They hate-watched. They doomscrolled. They lol-sobbed. They talked about "feeling seen," not always sure whether they were describing a blessing or a curse.

Surreality was standard. Instability was a way of life. In the world arising on the internet, content was farmed. Narratives were won. Orwell was a prophet. God was a mode. People coined new words. They made new memes. They built new canons. The art of earlier ages had its bloody wars and melting clocks; this one channeled its fears through a googly-eyed dog, animated but inert. It sits in a kitchen as the flames encroach, saying, "This is fine."

Whos or Whats

On the day of The Dress, I was #teamwhiteandgold—and for very good reason. I had spent much of that day as I do most days: screen-bound.

I grew up in the 1980s and '90s, in the era that brought the internet to the public but lived under the influence of television. "Screen time," back then, was a TV-focused anxiety. And I was, statistically, one of the kids the critics worried for when they warned about TV's influence. I loved TV. I watched it as much as I could. But I didn't lose myself to the screen, I don't think. In many ways, I found myself in it. I was raised in a coastal town with a rural spirit. My elementary school class had around twenty kids in it. My high school class had around ninety. Life, in the good ways and the bad, was small. TV made it just a little bit bigger. That glowing screen was entertainment, yes, but it was also travel. It was wonder. It was connection. It was education by other means. I loved the bigness of it, the possibilities it held—the daily evidence of the wide, weird world that stretched beyond the screen.

My fascination never faded. I'm a journalist now, writing about culture—which very often means writing about TV. It also means, very often, writing about the internet and the collisions it has fostered: between entertainment and politics, technology and media, pop culture and culture of a broader bent. The work includes stories about scripted series and reality, about podcasts and influencers, about information and misinformation—things I love to think about, but things I've found, in recent years, steadily more challenging to discuss. I kept finding myself caught in my own form of web-borne uncertainty: I could never seem to find quite the right words. And when you can't find the right words, it can be hard to find much else.

The people on the screen were inventing new categories and rejecting many of the old ones. Reality stars were actors who had been cast in the role of themselves. Podcasters were acting like journalists, sometimes, and sometimes influencers, educators, entertainers, propagandists. The categories blend and blur: Politicians act like reality stars; reality stars act like politicians; everyone is a performer and personality and brand.

None of that, on its own, was a problem. On the contrary, the collisions were evidence of the hybrid capabilities of the internet, working as they should. But they also seemed to be creating confusion about matters that used to be straightforward: blunt questions about what people owed to, and might expect from, one another as they shared the unsteady new stage. "Politician" is not just a job description, but also a set of behavioral—and ethical—expectations. A journalist has responsibilities that an influencer might not. My uncertainties, I began to realize, went well beyond matters of terminology. They were matters of grammar. What are we, precisely, on our screens? Are we facts or fictions? Are we *who*s, or are we *what*s?

On screens, we are both and neither, double and nothing. I wasn't sure how to convey the duality. Sometimes, when I'd write about the very real people who appear on our screens—reality stars, say—I'd catch myself defaulting to the show: I'd refer to them as "characters."

Logistically, this was a small problem, and easy to fix: I'd swap in a more accurate term ("cast members," maybe, or "performers," or simply "people") and get on with things. But the mistake kept nagging at me—in part because I kept making it, but also because, as small as it was, it wasn't minor. The difference between "character" and "person" is also the difference between . . . pretty much everything else. One is a fiction; the other is a fact. One is owed the respect of personhood; the other is owed nothing at all. One might hear something said about them—or read something, maybe, that a journalist wrote—and be hurt (or amused, or flattered, or indignant). The other, enjoying the sole benefit of nonexistence, is impervious to pain.

I knew all that. Still, though, I kept reflexively "character"-izing people, whether reality stars or content creators or podcasters or influencers or the everyday people whose traumas have been true-crimed into entertainment. I started noticing others doing the same. I began to wonder whether "character," the very wrong word, might also be, somehow, the right one: a mistake that captured something all too true about the people we know through our screens. Screens' alchemies had turned a dress into The Dress; maybe they were doing something similar to us. Maybe their light was skewing our vision. Maybe it was leading us to look at those most fundamental of facts—other people, warm and real—and see nothing but empty fictions.

Screen People rose from those uncertainties. Something was changing. I knew it in the way so many people know things these days—it *felt* true—and because I'd seen the evidence every day. But I wasn't quite sure how to talk about the shifts. I wasn't quite sure how to *think* about them. I could never seem to find the right words. And when you can't find the right words, it can be hard to find anything else.

I wanted to know what had happened. I wanted to understand why. I wanted to be optimistic. I wanted to know why optimism itself so often seemed consigned to the white-and-gold: You could find it if you looked. But you might be seeing something that, in the bigger picture, wasn't really there.

In 1951, the Mexican philosopher Emilio Uranga published a wide-ranging essay that attempted to describe the shared experience of being Mexican. As part of the effort, Uranga summoned philosophy and psychology and history and current events, considering Mexico's colonial past and its ever-evolving relationship with the world power to its north. The conclusion he came to was that the unifying feature of the Mexican identity was fracture. To be Mexican in the mid-twentieth century, Uranga argued, was to navigate endless uncertainty. It was to live within a "mode of being that incessantly oscillates between two possibilities, between two affects, without knowing which one of those to depend on."

The unsteadiness, Uranga suggested, was a cultural condition that became, inevitably, a spiritual one. The uncertainty would give way to melancholy. In it, "the soul suffers." It "feels torn and wounded." Uranga described the condition using the Spanish term for angst: *zozobra*.

The internet has turned zozobra into a technological condition, too. Screens, environments that are becoming ever less optional, consign us to ambient uncertainty. The video we come across on Instagram, we know, might be genuine, or it might be a deepfake. The image we're looking at might be doctored. The new friend we're chatting with might be a person, warm and real. She might be a character playing a part. She might be a bot. True or false, fact or fiction: The distinctions dissolve inside the screen.

The ambiguities, though born of the screen, rarely remain screenbound. They leap, instead, into the world at large. A few years ago, Monica Lewinsky, the one-time White House intern and a person who is more familiar than most with the consequences of being screened, shared an experience with her followers on Twitter. Checking into a hotel, she wrote, the person at the counter asked her, "Are you related to the actress Lewinsky?"

The clerk's question, like so many others these days, was both deluded and insightful: Lewinsky the person *was* related to Lewinsky the performer. Screens make everyone an actor. They make everyone an audience. We acknowledge as much in our scripts. We cast people as life's "main characters," sometimes to praise them and sometimes to mock them. We describe our own character arcs (and journeys, and eras). We live our own plot twists. Those who are misguided have "lost the plot." Those who are tarnished are "canceled." In earlier ages, people attributed their circumstances to the will of gods and the whims of fate. We attribute ours to the artistic choices of "the writers."

The general assumption—a world beholden to "main character energy"—isn't new. "Your story should be a movie" has long been one of the highest compliments Americans know how to give. We have long taken for granted that people are most admirable when they are "larger than life" and that events are at their best when they are at their most cinematic. An existence so full that it gets optioned—this is the American dream in its final form.

In the age of the screen, though, the language of the show has taken on a striking new literalism. "No notes," we might say, bringing the praise of the Hollywood producer to an outfit or meal or joke. (*"Iconic,"* we might add, as a flourish.) People now "soft-launch" new romantic relationships on social media, treating their followers like focus groups and their feeds as daily acts of PR. If the new relationships test well, they might be "hard-launched" and/or made "Instagram official": handles tagged, heart emojis deployed, captions crafted, likes sought and won.

The language is revealing—not because it's used literally, but precisely because it is not. "Hard launch" winks, just a bit: Love is not, as yet, a full-fledged consumer product. But love, too, is bending to the shape of the screen. Our hearts now have data trails. They can be broken with a swipe. Screens do more than host life's assorted shows; they demand them. They make performance inescapable. Romantic love has always been a public presentation as well as a personal feeling; now, though, it can be a piece of media. It can be something people "do for the 'gram." It can be content. It can be data. And because of all that, it can be subject to cynicism: Is it a performance or not? Is it real or not?

Language that catches on at scale tends to reveal resonant truths. Language that catches on so completely that it becomes cliché, meanwhile, might hint at resonant longings. "Authenticity" gained currency during a time of rampant fakery. The desire to act "intentionally" suggests all the things we can't control. "Empathy," too, is a revelation in reverse: a plea, issued into environments of casual cruelty. "Vibes" and "chaos" and "surreality" work similarly. They live in life's negative spaces. They are descriptions that acknowledge how many things have become indescribable.

Paradigm shifts, as the science historian Thomas Kuhn described them, are at their heart matters of language. Kuhn coined the term in the 1960s, after studying the theories of Copernicus and Darwin and Einstein—seismic shifts that have never fully stopped rumbling. Scientific revolutions can work a bit like political revolutions do. They can be disorienting. They can be wrenching. They arise when the old explanations can no longer contain the new truths. To live within them is to navigate constant doubt. It is to never be sure whether the dress is white and gold or black and blue.

The Endless Theater

Movie theaters, with their wall-high screens and tidy seas of seats, assume—and enforce—passive spectatorship. You are there as an audience member, nothing more. Dramas may play out before your eyes; couples may have their meet-cutes and their miscommunications and their climactic reunions; lives may be threatened; people may feel pain; they may cry out for help. All you will do, though—all you *can* do—is watch. You would be foolish, and rude, to do anything else. Bystander apathy in the moral universe of the theater is not a failing. It is a demand. Screens offer fantasies, and fantasies offer absolution: "Should I do something to help?" is a delusional thing to ask when the person in need is a trick of light, shimmering out from the projection booth.

The terms of all this are so widely understood that they are typically simply implied. Of course the screen people aren't real. Of course they're figments. But the tidy terms of the one-way screen become far less tidy when the screen is split in two. The light that flickers before our eyes might convey, still, fantasy and fiction. It might also convey people, distant but real. It can be difficult to tell the difference. Fourth walls are breaking. The stage is becoming endless. The old scripts are adapting to the assumptions of the infinite scroll. On our screens, we are both actors and audiences, both showrunners and extras, sometimes the stars and sometimes the scenery. We are cast and recast, as the show goes on. We might be writing the scripts. We might be acting them out. We might be directors. We might be spectators. We will be facts. But we may be seen as fictions.

We are only beginning to understand the consequences of that shift. But it is transforming our lives, bit by bit and pixel by pixel. Embodiment, as a mode of interaction, is becoming an exception rather than a rule. That, itself, is transformative. Our brains are primed for in-person interaction, adept both at reading other people and reading the room. We are less equipped to understand one another through the cold distance of the screen.

"We" is often not the right word: It can imply a commonality to people's experiences that isn't really there. But it is, where the digital world is concerned, the only one that works. We may each have our own little corners of the internet, spaces imposed and chosen and uniquely ours. But they are part of a shared and ever-expanding ecosystem. They are places where we might spend time, and very often places where we must. We live, in one way, as we always did, in a world of warmth and scent and breeze. But we also live within the screens. They are our habitats. They are our habits. They are our shared environments. Because of the internet's scale—because it is generally unrooted from the hard facts of the ground—its environs are places where words and images have outsized power to shape our sense of the world and of one another. And it is a place that is also a theater: a place of stars and bit players, of expendable characters and easy scripts—a place that prefers its facts to be convenient, dramatic, and molded in the service of the story. Seen through its screens, a world that is profoundly complicated can begin to seem, on the contrary, easy. The facts that are hard can begin to seem impolite, inconvenient, and wrong.

"Another Season of Coronavirus"

When 1984 drew to a close, Americans marked the new year with more than the usual celebration. The dystopia that George Orwell had imagined the year might bring—the brutal regime, the bleak tyranny, the language that curtailed thinking rather than expanding its possibilities—had not come for them. The Cold War was thawing; consumerism was booming; Skynets and Death Stars and Borgs, the Big Brothers of the new age, were safely relegated to the realm of fiction. The year 1984 had not, for all practical purposes, become *1984*.

The scholar and cultural critic Neil Postman acknowledged the milestone at the start of his 1985 book *Amusing Ourselves to Death*. And then he broke the bad news. People had been measuring themselves against the wrong dystopia, he argued. It wasn't *1984* that had the most to say about the America of the 1980s. It was Aldous Huxley's *Brave New World*. "In Huxley's vision," Postman wrote, "no Big Brother is required to deprive people of their autonomy, maturity, and history." Instead, "people will come to love their oppression, to adore the technologies that undo their capacities to think."

The vehicle of their oppression, in Postman's mind, was the device that so many Americans associated with their freedom: the television. The device, Postman argued, had thoroughly insinuated itself into all elements of American life—and not just in the boob-tubed, couch-potatoed, the-average-American-watches-five-hours-of-television-a-day kind of way that was so common a criticism at the time. Instead, for Postman, TV was a technology that doubled as a value system. It did not merely entertain people; it shaped them. It made them restless. It made them distractible. It made them believe, after a while, that life was at its best—and its most believable—when it resembled a TV show. Politics, education, science, religion—they had all twisted, in their own ways, to TV's demands. They had learned, Postman thought, that medium's most elemental message: If something was to be taken seriously, it would need to be entertaining.

Because Postman criticized TV (and because he took on "amusement" in the title of his most famous book), he is sometimes remembered as a scold: humorless, judgmental, chiding people for their fun. But *Amusing Ourselves to Death* was a clever act of bait-and-switch: It is a book, in the end, about technology. Postman had studied—and was deeply influenced by—Marshall McLuhan. As he made his case for the dangers of television, he considered the history of earlier communications technologies: the printing press, the telegraph, the written word itself. "The medium is the message," for Postman, was more than an axiom; it was a warning. It was a reminder that our mediums, if we're not careful, can become our mindsets. "The problem is not that television presents us with entertaining subject matter," Postman wrote, "but that all subject matter is presented as entertaining." Watch enough TV, in other words, and soon enough you'll expect the world to act like a TV show.

It might seem, today, that Postman's criticisms have aged into obsolescence—arguments made archaic by the age of the ambient screen. The arrival of the internet, though, has only brought more acuity to his insights.

Screens, precisely because of their two-way workings, make the living of life and the performance of it ever more indistinguishable. On screens—and through them, and within them—we watch the spectacles. We are the spectacles. We become one another's fun. We defer to entertainment as a value system.

"There's this mindset that it's like running a show and you've got to keep people tuned in, you've got to keep them interested, and at some point you've got to move on and move on quickly," a former Health and Human Services official told *Politico*, discussing the Covid pandemic. "Viewers will get tired of another season of coronavirus."

It's a worry that would be echoed by many other officials as the pandemic stretched on from weeks to months to years. But this particular concern about national channel-changing was notable mostly for its timing. It was expressed in May of 2020—the early stretches of the Covid crisis in which so many people were dying that New York City had to bring portable freezers to hold the corpses of those who had succumbed to the disease. If Covid was a TV show, it was at that point a medical drama—and one whose every episode was primed for the demands of sweeps week.

Fiction is creeping into every corner of life. Americans have grown accustomed to having even our most urgent and high-stakes tragedies fed to us as pieces of entertainment. We are used to understanding the world through the structures of the TV show: the changing plotlines, the exciting new characters, the expendable characters, the acts of winking fan service.

Soon after Russia's attack on Ukraine, *Vanity Fair* wrote an article describing the resurrection of an anxiety that seemed, at one point, to have been consigned to history: the possibility that the war might lead, in short order, to a Russian attack on the United States. The magazine described the threat, overall, as "the gritty reboot of Gen X's nuclear nightmares." In March of 2022, the digital news site *The Recount* shared the news that researchers had identified a new variant of the coronavirus: "COVID's omicron spinoff now has its own spinoff."

Semantics, you might object: Comparing Covid to a show is not the same as believing it to be one. The demands of the show, once they infiltrate things, have a way of settling into people's expectations of what life is, and should be. Americans are learning about current events through podcasts and TV shows and movies and clips that have more incentive to be interesting than to be true. They are getting medical advice not from doctors but from social-media influencers. They are getting information that way, too.

Real people's traumas—murders, kidnappings, rapes—turn into other people's amusements. Many people, in the choose-your-own-adventure environment we are creating, are coming to resent facts. Basic information is struggling for viability against the lies. Conspiracism, and the cynicism that fuels it, is gaining purchase in American life. "Crisis actors" are go-to scapegoats not only of conspiracy theorists but also of the growing number of people who treat the truth as an inconvenience. Cruelty is seeping into the warmest corners of people's lives. Too often, the spectacle and the suffering are no longer opposed to each other; instead, they become indistinguishable.

This is what you might call fiction creep. Disguised as low-stakes fun, fiction—fun, endless amusement—is stretching into an expectation. Many Americans, after all, *did* tire of Covid, the show. As the days of Covid stretched into weeks, and months, and years, people compared the whole thing to *Groundhog Day*, resenting the stultifying sameness of it all. They complained about the virus's failure to adhere to a tidy plot arc. They would find other sources of drama in those days of "lockdown" and social distancing—more on those soon—but the side dramas were necessary, apparently, because the overall show had lost its edge. If a global pandemic is going to drag on, it should at least offer some decent plot twists.

Covid brought a grim new currency to Stalin's insight that "a single death is a tragedy, and a million deaths are a statistic." When the pandemic has an audience to please, a single death becomes a plot twist. A million deaths, though, begins to look like lazy writing.

Spectacle is our medium and our mandate: We become so accustomed to the performance—the sparkle, the glitter, the drama—that the unscreened world comes to seem dull by comparison. We now risk being unable to process our facts through anything *except* a show. When you walk the world wearing screen-colored glasses, it's easy to forget that the vision is tinted. It's easy to assume that the world itself is as shiny as the lenses suggest. And it's easy to demand that it be.

All the world's a stage is steadily transforming from a metaphor into a mandate. Where does the performance end and life begin? The answer has never been straightforward, but it has also never been more muddled. If you assume that the world exists for your entertainment, it also becomes easy to assume that the world owes you the same thing entertainment does: ease, delight, maybe a catharsis or two. You might start to expect that history, too, should be easy. You might start to feel the desire, when confronted with stories that are not soothing or tidy, to scroll over to a different screen.

Screens give; they take away. They can encourage connections between people; they can foreclose them. To live among screens is to be unsure, always, where the performance ends and the reality begins. It is to know that the video we come across on Instagram might be genuine or a deepfake; that the image we're looking at might reflect something real or distort it; that the new friend we're chatting with might be a person, warm and real, or an actor playing a part, or a bot, scripted into seductive humanity.

In the land of the screen, anything might be transformed into low-stakes amusement. History might become a rollicking tale that is merely "inspired by true events." When it fails to entertain—when it fails to comfort or inspire or provide a tidy ending—it might offend. After a while, the news that is dramatic and the news that is important come to look like the same thing. The emergency becomes soothing. The spectacle becomes beguiling. We get so accustomed to the hyperdramatic environment that the plain old version starts to seem like a disappointment. Like the untouched photo next to the filtered one, the world itself—its plain facts and muted hues—comes to look like footage merely waiting to be edited.

After a while, drama (and crisis and emergency) becomes our default state. We do not merely use it as one mode among many; instead, we rely on it. We come to see the drama as the only way to make change. The best time—the only time—to talk about gun-safety legislation is during the window cracked open by the latest mass murder. The time to talk about climate change is just after the latest disaster.

We wait for the world to happen to us and then for the happening to be converted into a compelling show. Only then do we act. If, indeed, we act at all. Shows are seductive. They are easy. They are exciting. Entertainment creates audiences. It does not, however, create publics.

Screen People

Elizabeth Eisenstein, the great historian of the printing press, was also a scholar of technological revolutions. Their upheavals, she argued, do not tend to arrive with the invention of a revolutionary technology. They come instead in the aftermath of the invention—gradually and then, quite often, suddenly—as the new technology spreads and normalizes. New technologies, to paraphrase the web theorist Clay Shirky, don't become interesting until they become boring. This is how it was with the printing press: Its full impact would not become clear until more and more people would experiment with it, live with it, profit from it, reckon with it. A technological revolution, in that way, doubles as something of a paradox: Only once the new tool reaches a level of stability in a culture will it destabilize everything else.

We are in the early stages of our own revolution. Through the internet, the physical world is colliding with the disembodied realms of the screen. The old things are breaking; new things are emerging, haphazardly, in their place. Categories that once offered order are collapsing; new ones have yet to be defined. Our addled language reflects our addled lives: the confusion, the exhilaration, the exhaustion. The world can feel chaotic right now because it *is* chaotic.

Revolutions, whatever their form, are rarely pleasant for the people who live through them. They can break people's faith, and, with it, the bonds that once held them together. They can engender widespread uncertainty, and all the ills—anxiety, fear, suspicion—that can come in its wake. Life will always be a tug-of-war between the known past and the unknown future; to live in this moment, though, is to be especially torn between what has been and what might be. There are few certainties in this in-between age.

Instead, our two-way screens consign us to confusion. When even the most elemental of distinctions—fact or fiction? person or character?—fall away, everything else becomes just a little more unsteady. "A man with a watch knows what time it is," the saying goes. "A man with two watches is never sure."

The chapters to come, I hope, will provide a bit of steadiness. They draw from work I've done at *The Atlantic* and from the research I've done as I've tried to explain things to myself. The stories they tell are not always happy, but they are hopeful. The internet, though it is hardening, has not yet solidified. We still have time to shape it. We can decide, still, how we want to be human together. That task is unprecedented, but it is not wholly new. People have navigated technological revolutions before. They have come out on the other side, changed but intact. We can do the same. "Anxiety," Søren Kierkegaard wrote, "is the dizziness of freedom." If so, it seems a price worth paying. We can write new scripts. We can update the old ones. We can try our best to see one another as we are. All the world's a stage—until it isn't.

1
The Screens

An early moment of *It's a Wonderful Life* takes place at a high school dance. The teenagers are in their school gym, clad in suits and frilly dresses, swinging around the floor. The scene is set in the 1920s, but it is, in another way, timeless. You can almost feel the heat of the room, the thrum of the music, the joy of all the dancing.

But the gym's floor, it turns out, is modular—laid over an indoor swimming pool. All it takes to retract the floor is a simple turn of a key. As his classmates have their fun, a prankster sees his chance. He twists the key. The floor the students are dancing on begins receding. George and Mary, the film's central couple, end up near the edge. The pair, so focused on each other that they're oblivious to everything else, dance themselves off the floor, falling into the water.

The prank does not end the good time. Instead, the revelers begin leaping into the water, turning the dance into a pool party—the kind of thing they might tell their grandkids about one day to show they were young once, too. But the pool scene is also a hinge point, for George and Mary and many others: It is a last gasp of innocence before their world begins falling apart. Soon after the dance, George's dad dies. The market crashes. War comes. *It's a Wonderful Life*, that unlikeliest of holiday classics, makes instability so complete that it becomes an environmental matter. The ground in Bedford Falls is never solid. George and his friends are sledding across the pond; the ice collapses below. The young people dance, too caught up in their fun to notice that the floor has been pulled out from under them.

That scene, nearly a century removed from our moment, anticipated something of it, as well. The dancers' lot is ours, too: Things that were solid are cracking, giving way to the liquid below. The internet rushes forward in feeds and flows, in streams and infinite scrolls. Things blend. They bleed. They collide and sometimes combine. The hard-edged mediums—video, audio, text, more—live within it, still, but they also collide, sometimes seamlessly and sometimes awkwardly. Online, people talk through text. They laugh through icons. They trade the old categories of knowledge, facts and vetted information, for hazier ones that make few distinctions: a world constructed of "content."

Along the way, the old assumptions—the frameworks that once structured our lives—are losing their footing. Categories are collapsing. Shared truths are dissolving. The cultural theorist Lauren Berlant, writing in recent years, described the "genres" of everyday life: the rubrics people rely on to organize the bigger stories they tell themselves about the world and their place within it. The genres themselves aren't necessarily fixed and can vary from person to person. But they can ground people, as individuals and as communities. They can help us to make sense of ourselves.

Set against the liquid internet, though, the old categories begin to erode. If we're not sure whether the people on our screens are *who*s or *what*s, the other genres fall away. "From falsehood, anything follows," says an ancient law of logic. It applies to untruths of the most foundational kind: the distinction between fact and fiction.

Genres have moral grammars, too. Comedy is different from tragedy not just in its form, but in the assumptions it makes—and the guidance it gives—about the way audiences should react to it. Say someone slips on a banana peel in a comedy, and say the same thing happens in a drama. Very likely, you'd laugh at the first scene and wince at the second. And the difference would come down to the assumptions about empathy that help to distinguish the drama from the comedy: In a comedy, generally, audiences understand the banana-slipper as a character. In a drama, generally, they understand her as a person.

On interactive screens, though, genres collapse, as well. Everything becomes a matter of both and neither, double and nothing. This is the core fact of our new medium. It is the root of many of our problems. Two-way screens separate things that had been connected. They blend things that had been distinct. In the internet's flow, things that are united break apart. Things that are separate collide. Tragedy and comedy, fact and fiction, the living of life and the performing of it: They flow so steadily into the single stream that they become ever harder to disentangle.

No wonder "chaotic" has become a default description for everything from people to world events to that sweater in the window at Zara. Or that "vibe," with its you-know-it-when-you-feel-it elasticity, has become similarly ubiquitous. Or that so much, these days, is "surreal" or "unreal" or "unhinged" or "bonkers." Dictionaries naming their recent words of the year—terms meant to convey something deep and true about The Way We Live Now—have emphasized the language of web-flavored fracture: "hallucinate," "post-truth," "gaslight," and, courtesy of the American Dialect Society, "enshittification." Oxford's 2024 selection, "brain rot," offered a mass diagnosis of an illness with no cure.

The words capture something essential about the instabilities of our heady age. And I've found them indispensable, as a writer and as a person. I've described a range of things (news events; a salad I've made; myself) as "chaotic." But "chaos" and its similar terms are useful precisely because they decline to describe much of anything at all. Instead, they acknowledge the broader truth: all the things that can seem, in this seismic moment, indescribable. As the ground gives way beneath them, they leap into the pool.

In 2021, crowds in London came out to celebrate Queen Elizabeth II's Platinum Jubilee, the event marking the seventieth year of the British monarch's reign. The traditional pageantry involved the procession of an ornate coach through the streets of London. A royal coach procession is a lengthy affair, though, and the ninety-six-year-old monarch had elected to preserve her energy for alternate events in the celebration. But this show, too, would go on. Inside the vehicle, then, in lieu of the queen herself, was "the queen" herself: a hologram that waved with robotic precision at the crowd.

The whole thing was fairly standard, as surreality goes. What was most remarkable, as the whole process played out, was not the queen's ethereal body double. It was that the crowd, as the queen waved at them, waved back.

People describe the world as "surreal" for good reason: Things *are* surreal. Bots flirt with us. Pictures lie to us. Regal holograms wave to us as they pass by in their gilded carriages. Monarchy claims that some people can be human and inhuman at once—but Her Royal Hologram, greeting her subjects, brought a new literalism to the premise. The people who waved to her were not foolish; they were aware, for the most part, that the queen they saw was the "queen." They were simply caught between two paradigms. They had no protocol to guide them in the presence of a monarch who had no presence at all. So they applied the old standards to the new reality: When the "queen" waved at them, they waved back.

Words, Written and Said

Screens, so suited to the transmission of images, have also brought a new primacy to that ancient medium: the written word. The many transformations that have been brought by video-based communications platforms have been mere accompaniments to platforms that allow people to write to each other. The most widely used social-media apps, by a large margin, are the most basic: WhatsApp, iChat, and other services whose primary offering is . . . text. Much of the content shared on even self-consciously image-oriented apps—Instagram, TikTok, and the like—are captions and comments delivered through words (and, sometimes, emojis).

The rise of text in this age of video may be a slight irony, but it is consonant with the particular kind of chaos that technological revolutions can bring. The printing press ushered in changes that would amount to a new culture and a new age: the age of print. Even as those changes were taking shape, though, Elizabeth Eisenstein observes, the new technology was also ushering in a new age of imagery. Challenging the long-held assumption that print outmoded images and illustrations, Eisenstein argues that print, in fact, restored old forms of iconography that had been lost to centuries' worth of scribe-created texts. Through woodcutting and similar image-replication techniques, images could be mass-produced in ways never before possible. They could spread the way words could.

Eisenstein also challenges another widely held assumption. "It was not the printed word but the printed image that acted as a savior for Western science," she argues. Woodcuts, which brought the ink-and-stamp logic of movable type to picture-making, also brought a new specificity to maps and to depictions of nature more broadly. They allowed for the creation of shareable tables, diagrams, medical renderings, and drawings of newly discovered plants, animals, and geological forms, revolutionizing the study of geography, astrology, botany, anatomy, math, and many other subjects.

The scientific images were accompanied by mass-market illustrations, from cartoons to depictions of Biblical stories—opening a world of knowledge and entertainment to the many people who could not read. The woodcut images introduced the logic of the photograph—"The exactly repeatable pictorial statement," in the scholar William Ivins's term—centuries before the invention of the camera.

The web has brought a similar kind of ferment. The words are proliferating even as the images are. But the two forms are not merely coevolving. They are also collapsing into each other. We are losing the old distinction between written words and spoken ones.

The changes brought by the introduction of the printing press affected every area of culture—in part because they were matters of language. The printing press steadily changed how people communicated, and that, in turn, changed everything else. The process was not tidy, but it was transformative. Before the printing press, Europe was an oral culture, shaped by words that were heard rather than recorded, its literature shaped around the demands of memory—repetition, rhyme, emotionally resonant imagery—its sense of the past limited to the lore.

Oral cultures tend to value the individual ability to memorize information more than print cultures do. They also tend to prize—and celebrate—the communal elements of knowledge-making: Each person bears the dignity and the responsibility of the collective memory.

Take, for example, *The Iliad* and *The Odyssey*, epic poems that weave fantasy and history, turning the past into a rollicking adventure inflected with magical realism. The flourishes that give the stories their poetry ("the heat of Love," the "wine-dark sea") were also mnemonic devices: Listeners, feeling the words as well as hearing them, were more likely to remember them. Though "Homer" is cited as the poems' author, he was most likely not one person but many: bards who passed the stories down, line by line, scene by scene, perhaps adding their own flourishes along the way. The stories were long ago written down—scrawled on papyrus and parchment, captured by the scribes—but they speak of a time when words, once uttered, slipped away.

Writing, that elemental technology, made language newly powerful: Words, remade as objects, could outlive their human agents. The printing press expanded that power, and made it newly personal. Even in its earliest days, print demanded—and, thus, encouraged—new ways of thinking. Whether people were processing linear text or parsing diagrams, the "brain work" required of them, in Eisenstein's term, was much different than the work required to make sense of words that, as soon as they were spoken, gave over to silence.

We are heirs to the paradigm that arose as, day by day and brain by brain, people adjusted to words that were written and organized and standardized and widely shared. We have lived, as Marshall McLuhan—and, later, Neil Postman—suggested, in a print culture. We navigate the world with brains whose synapses and circuits have conformed to the demands of text.

But we do so using technologies whose cultures are neither print nor oral. Instead, they are rooted in both cultures at once. A platform like X, for example—the short-form publication service previously known as Twitter—is primarily text-based but also guides users into the kind of quick-twitch reactions typically reserved for spoken conversations. Statements made on the platform reflect the ephemerality of words uttered orally—posts pop up for a moment, only to disappear into the infinite scroll—but are also preserved indefinitely. "The internet never forgets," the warning goes, and its steel-trap memory makes no distinction between words offered in the spirit of writing and those offered in the spirit of speech.

"How to talk" and "how to write" are related propositions but fundamentally different ones. The distinction between them is, in general, so deeply ingrained that we are no longer conscious of it: You use the one for memo-writing or letter-writing or book-writing, and the other for . . . most of your other daily interactions. There are shades of gray, certainly (a letter sent to a friend might be peppered with chatty nonchalance, while a conversation with your boss, you might decide, calls for a bit of the letter-writer's formality). For the most part, though, we are caught in a world-historical plot twist: We've adapted ourselves to a technological environment that is rapidly ceasing to exist.

Facts, as a category of knowledge, did not simply happen. They had to be created, argued for, protected, and widely acknowledged as superior to truths that were informed by faith and emotion alone. And the values they engendered—the broad notion that the world should be shaped by reason, logic, evidence—are relatively recent developments.

Facts, in other words, are contingent, and as such much more fragile than they might seem. And a culture that prioritizes the compelling stories over the true ones is a culture that puts an evidence-based approach to the world—which is to say, reality itself—at risk.

McLuhan and Postman, when they are remembered today, are often associated with their most sweeping maxims. McLuhan, describing the interplay of the machine and the information it shares, had *the medium is the message*. Two decades later, Postman gave the formulation the critic's version of a ringing endorsement: He suggested that it hadn't gone far enough.

For Postman, mediums were so consequential that they gave rise to paradigms. For much of American history, print had been the dominant medium and thus the dominant culture: People learned about the world—consumed it—through newspapers and books and pamphlets, their words arranged in tidy lines, their claims organized into coherence.

As a result, Postman argued, people naturally came to value the type of thinking that text required: logical, patient, evidence-oriented. And that value shaped everything else. "In a print culture," Postman observes, "writers make mistakes when they lie, contradict themselves, fail to support their generalizations, try to enforce illogical connections." In a print culture, essentially, you make a mistake by being wrong. In a culture shaped by television, on the other hand, you make a mistake by being boring.

Applying McLuhan's adage to American culture as it existed in 1985, Postman proposed a broader iteration: *The medium is the metaphor*, he argued. By that, he meant that a medium will impose itself on minds as well as messages. It will become an influence over people's lives—and then, soon enough, an expectation. And then, soon enough, a demand. To live in a TV-centric world, Postman argued, is to expect life itself to play out as a TV show—abidingly entertaining, narratively satisfying, passively consumed. If TV is an ideology, its values cohere to a single party line: TV's job is to make audiences. Audiences' job is to watch.

The arrival of the internet—first with its transition into a consumer platform in the 1990s, and even more so with the advent of social media in the early 2000s—has both validated Postman's ideas and, in other ways, complicated them. Television, now encompassing terrestrial channels, cable, satellites, and digital streams, still assumes passivity: People may consume its content while commuting or side-screening or doing the dishes, but they are, fundamentally, *consuming*. Many of the web's TV-like extensions, from YouTube and other video-sharing sites to scroll-oriented social-media apps, work in similar ways: They assume, even if they encourage engagement, a passive spectatorship. They present the content; the users watch the show.

The endless theater is changing us in ways that McLuhan and Postman both predicted and could never have imagined. It has thoroughly validated their maxims. It has also partially outmoded them. People, now, are the content on the screens. We are the shows on display. This is the core of our chaos. *The medium is the message*, McLuhan wrote. *The medium is the metaphor*, Postman countered. Interactive screens suggest a third iteration: *The medium is the moral*.

I came to this book as someone whose job has been transformed by the internet in nearly every way—but one way in particular. In the past, I loved corresponding with readers. Even when people wrote to disagree with something I'd written, they seemed interested in explaining why—in having, over email, a kind of conversation. In recent years, though, the notes have steadily . . . changed. Many come with invectives. Some come with death threats. Some of those are quite graphic. These are the notes that have made it through the spam filter; they are typically responding to things I've written about *Jeopardy!* or *Ted Lasso* or some other low-stakes show. "I don't agree," they seem to be saying. They simply choose to express the disagreement, to the stranger on the other side of the screen, as a wish for the stranger's demise.

My experience of that has been mild compared to those of many of my fellow journalists. And while I didn't relish receiving the notes—each seemed a new symptom of the same vague disorder—their primary effect was to spark my curiosity. Email had been around for ages; why were messages like this coming now, with such regularity? What had changed, exactly?

Screens, as environments, do what every environment will: They steadily influence their inhabitants. And social-media platforms, though distinct from the wider web, tend to influence everything else. The spaces they build up around people can drive them toward mindless extremity. The conversations they encourage can lead people to think that "I disagree" is best communicated as "You should die."

Reading the Room

Frank Capra, the director of *It's a Wonderful Life*, once revealed one of his favorite filmmaking techniques. "I like to speed up the pace of my films," he wrote. "The pace of my films is much faster than normal, because I think things slow down on a large screen." He introduced the speed, he suggested, not because of the demands of film—but because of the demands of an audience:

> I don't know if stimuli affect people more quickly because a thousand pairs of eyes and ears are perceiving something simultaneously, or if those thousand pairs of eyes and ears simply accept stimuli faster than we think. But I do know that a person in a crowd reacts faster to images than a single person would. I know that you have to speed film up to make it look natural to a thousand people.

Capra, that bard of American individualism, was identifying a principle of the science sometimes known as "deindividuation": Crowds are not merely collections of people; they are collections that change people. The pioneering sociologist Émile Durkheim described the "social current" that can sweep gathered individuals along in its flow. "In a public gathering," Durkheim observed, "the great waves of enthusiasm, indignation, and pity that are produced have their seat in no one individual consciousness. They come to each one of us from outside and can sweep us along in spite of ourselves."

Crowds, that is to say, are forces of physics. Their mass, itself, propels them. And the propulsion exists even in the disembodied crowds that gather in digital spaces. When individual people become a collective—when they are crowded together, as audiences or participants or something in between—they transform. They might seem like data sets. They might seem like "mobs."

Online, the alchemy Durkheim described can conflate the difference between "social currents" and what Durkheim called "social facts": the values that are externally held, collectively, but that exert their pressure on individuals. Humans are social animals, and this means that our sense of other people's opinions will very often come to shape our own. When people believe something to be a matter of consensus, scientists have found, they tend not only to go along with that idea in social settings. They also tend to internalize the "consensus" opinion as their own—for better or for worse.

When Americans talk about social interactions as ongoing performances, they do so, in part, because of Erving Goffman. The sociologist, through a series of books published in the mid-twentieth century, analyzed the dramas of everyday life, identifying the scripts—silent but always there—that seemed to govern those exchanges. Goffman's ideas were revolutionary in part because they were so obvious: He was taking stage directions that were commonly understood in the settings he studied—directions that also go by the name of "etiquette," "culture," and the like—and summoning them out of their silence. He was making the implicit explicit. We read each other's behaviors like texts, essentially, and respond in ways that will be read by our scene partners. Social engagement requires constant acts of microanalysis and extrapolations, as we read the performances to make broader assumptions about the people we encounter.

Goffman was taking the stuff of frustration and poetry and art—that we are always mysteries to one another, in some way, and always trying to solve one another—and turning it into a science. In many ways, even his parsings were operating as metaphors: The scripts that govern social interactions are not literal. They don't need to be. They are simply commonly understood.

But Goffman's theories did assume one basic kind of literalism: The dramas would play out in person. The "presentation of the self," as Goffman called it, was achieved through physical interactions, within physical spaces. When he talked about the front stage and the backstage—the dramaturgical divisions, as he saw them, between private places and public ones—he quite literally meant the physical demarcations between, say, the home and the public sphere. His insights about endless performance took that defining feature of human biology—our brains, honed over millennia to serve us as social animals—and applied it to human society. In person, consciously or otherwise, we are primed to detect our scene partners' tiniest flinches and flashes of emotion. We're great at reading each other. We're great at reading the room.

When human dramas go digital, though, many of the scripts Goffman identified—all those tacit stage directions—can fail. Reading one another across the distance of the screen is a new kind of challenge when the people we are trying to make sense of are flat images. Life's scene partners, in Goffman's analysis, do not merely perform the scripts together; they also collaborate, as he put it, in "fostering a given definition of the situation." As people interact according to the existing scripts, they also create new ones. That shared, tacit, real-time scripting is a delicate balance of power—not just between individuals but between each individual and the norms that shape their cultures. Goffman's theories, in that sense, anticipated the tensions of the screen. Scripts connect people. They constrain them. They make performance inescapable.

Marshall McLuhan, considering the stage itself, foresaw similar tensions. Televisions, by the mid-twentieth century, had already nullified distance and time, allowing moving images to be transmitted from one location to another, recorded in one moment and broadcast into another. The machines of the future, McLuhan believed, would continue that work. They would expand life, in some ways. They would shrink it in others. They would, over time, steadily remake the world. They would give rise, McLuhan argued, to a "global village."

McLuhan offered the assessment shortly before universities in California would link their computers through a single network—and roughly two decades before the US Defense Department would develop a protocol that allowed for inter-network connections. In some sense, then, McLuhan foresaw the internet. But his most enduring prophesies—and the insights that have retained their urgency as the internet has shifted from an idea to a banality—are not, directly, matters of machinery. Instead, they are matters of culture.

"The global village" is sometimes read today as a place of transcendent humanity: a world where, distance having been obliterated, connection triumphs and empathy reigns. But the village as McLuhan presented it was no such democratized dreamscape. Yes, it would connect people, he thought. It would give them unprecedented access to one another. But it might also create new forms of distance. It might cause misunderstandings, envies, resentments. Villages can be inviting. They can also be stifling. They make people accountable to one another precisely because they also make people inescapable from one another. Proximity, like so much else, has a double edge: It can foster empathy. But it can also, at times, foreclose it.

"Humanity," as a species and as a broader designation, is a product of endless adaptation—to the physical world and to our knowledge of that world's workings. Galileo, his sight sharpened through the curves of new lenses, changed the story people told themselves about Earth's place in the cosmos. Darwin's sailing trips expanded his vision, too. Observing the planet's creatures from a new perspective, he saw patterns that allowed him to see a story: life, all of it, arising from and then connected by its drive to keep living. Natural selection, as a theory, was at once elegant and primal. It was also vaguely heretical.

"Have dominion over the fish of the sea, the birds of the air, and all the living things that move upon the earth," Genesis had said at its outset, and the biblical line, in England and elsewhere, had grounded a broader form of faith. Humans, the assumption went, dominated not only by brute force, but also by divine right. Their exceptionalism as a species rationalized their rule.

The Origin of Species, published in 1859, told Earth's apex occupants that their status had been an illusion. They may have invented agriculture and art and civilization itself; they were animals all the same. The book itself is fairly apologetic on this score. Darwin, as a scientist, emphasizes the data he compiled and the details he observed on his trip to the Galapagos, offering it in meticulous detail. As an early Victorian gentleman, however, he allows the uncourteous consequences of his findings to go, for the most part, unsaid. *The Origin of Species* was science that might have been read as melodrama: *A man went bird-watching, and now we're all questioning God.* The book, as it presents its theory, conveys its preference that you not blame the messenger. It is the rare piece of scientific research that reads, also, as a euphemism. But the subtext was clear: The science, here, was personal. Darwin, in the name of human knowledge, was making humanity itself newly questionable.

But that is, in some ways, a common transaction. Some of the earliest known forms of writing, the author James Gleick suggests in his 2011 book *The Information*, were legal documents: charters and deeds executed in dutiful protolegalese. But the writers themselves, operating from an oral paradigm that understood contracts to be conversations, would sometimes add discursive flourishes to the documents. "Oh! All ye who shall have heard this and have seen!" one said. Several charters ended with a polite valediction offered to their readers: "Goodbye."

Their confusion is so familiar. It is the same kind of category error people made in the early days of texting, when some would begin messages with a polite "Dear SoAndSo" and end them with formal sign-offs. None of that was wrong. It was simply not reflective of the new medium's form and function. Like those ancient scribes, we are bringing old brains into new tasks. Awkwardness along the way is inevitable.

The scribes were making rational assumptions: Their words would have an audience, and the reasonable thing to do with an audience is to acknowledge them. But writing, like every communications technology, will impose its own sense of the rational on its human users. "Before writing could feel natural in itself," Gleick writes—and before it "could become second nature—these echoes of voices had to fade away. Writing in and of itself had to reshape human consciousness."

Humans of today are in the midst of a similar reshaping. The difference now is that the technology that molds our minds is rooted not in a single paradigm but in two at once. The advent of the radio brought a form of "re-oralization," many scholars argue: Words aired on the radio were both "embodied"—they could only be delivered by a human voice—and, for their audience, ephemeral. Television continued the process, with imagery as part of the package.

The cultural historian Walter Ong assessed the impact of both in his 1982 book *Orality and Literacy: The Technologizing of the Word*. In it, he outlined the now commonly invoked distinctions between oral cultures and print cultures. And he used the framework to argue that the new technologies were ushering in a new age of orality—one in which spoken words proliferated against the backdrop of the established print culture. He described the new situation as "secondary orality."

Ong's frameworks, many of them published well before 1982, had the kind of radiating impact most scholars can only dream of. They influenced the work of McLuhan, Postman, and many other thinkers. Today, they also inspire debate: Are the frameworks patronizing? Are they overdetermined? Or, perhaps, is their problem that they didn't go far enough?

Ong's theory was, in its way, diplomatic: A culture shaped by secondary orality is a culture that can benefit from both oral and print-based approaches to information. Recent years have given rise to a more Manichean proposition: the Gutenberg Parenthesis. Thomas Pettitt, a professor at the University of Southern Denmark, argues that the dynamics Ong described are giving way to a new historical era—an era that is returning us to our cultural past. The culture that developed in the centuries that followed the printing press, the idea goes—the period from, roughly, the fifteenth century to the twentieth—was essentially an interruption in the broader arc of human communication. The text-based paradigm that we currently inhabit will give way, in short order, to the oral communication methods that defined human discourse in the years before moveable type: conversation, gossip, the immediate prioritized over the enduring.

Both theories help to explain where we are. But neither takes the full reality of the web into account. The web's defiance of structure—all those feeds and flows and endless scrolls—exerts itself even on the distinction between spoken words and written ones. Oral culture won't overtake print culture. But nor will oral and print paradigms coexist with minimal disruption to the culture that was already there.

Instead, the paradigms will tangle and blend, creating new values and likely making many of the old ones obsolete. Facts will lose their centuries-in-the-making sanctities. Some information will be assessed according to its accuracy; some will be assessed according to its palatability. People will debate about which metric is more viable, and the debate, itself, will be evidence of a fight already lost. They will use language not merely to reflect the world as it is but also to will their preferred world into existence. Words will become more powerful, and therefore more confounding, than ever before. Americans talk a lot about post-truth politics, as we should. We should also, however, consider the consequences of lives lived as post-truth people.

McLuhan, after proposing the "global village" of the future, began to wonder whether his own term needed an edit. Satellites were steadily expanding television's influence. Nature was being remade as an ever more synthetic environment. The planet would still be subject,

McLuhan thought, to the distance and proximity of the "global village." But it would also be subject to the needs of programming. In 1970, McLuhan offered a new term for understanding how the world might be remade by screens. The "global village," he suggested, would also be the "global theater."

2
The Stars

In late 2006, *Time* magazine announced its latest Person of the Year: You. The choice was both pragmatic—"You control the Information Age," the magazine declared, by way of explanation—and symbolic. YouTube had launched the year before; Facebook, the year before that. The iPhone would come in 2007, making screens responsive to warmth and touch, and the internet newly companionate. Its many competitors followed. These were the years that found the standard web turning into the social web.

They were the years, as well, that were finding screens becoming ever more portable and ever more ubiquitous. They contained not only the internet, but also the phone, the voicemail recorder, the microphone, the alarm clock, the calculator. People used them to find employment, friends, love. They sought comfort in them when those things were lost. The phones held people's diaries. They held their photo albums. They held the last voicemails their grandfathers sent—"Just checking in to say hello, honey . . ."—before they passed. They held the data, and the data held lives. That one screen, when activated, was a portal to the digital world.

Time was acknowledging the shift. It was doing so extremely literally. Print versions of the issue replaced the typical honoree portrait with a nearly cover-wide image of a computer monitor. Inside its "screen" was a foil-like surface meant to mimic a mirror: You, elevated by the internet.

The whole thing was easy to write off: as a stunt, as a play for attention, as a ploy for newsstand sales when digital consumption was on the rise. But the gimmick—*Self-Portrait in a Mylar Mirror*—was, as *Time* explained it, stridently sincere. The internet, the magazine suggested, would bring democratization and, with it, a new era of prosaic dignities. And *Time* was not alone in its optimism. The internet, back then, doubled as an act of faith. It would bring people together, they thought. It would swing history on its hinges. In the glow of the screen, people would find brave new ways to be human.

The Information Age itself, to an extent, justified the optimism. The internet brought people together; the hyperlinks that bound its pages made bonds among humans, too. People watched one another, through the screens, as they always had and as they had never been able to before. They tagged one another. They viewed and commented, subscribed and liked. They gave out hearts as tender. They judged with thumbs and stars. They grinned through their keyboards. They talked through the text. They were expressing themselves. They were documenting themselves. They were doing something big and new.

Scholars, watching it all, talked excitedly about "participatory culture" and the potential of society-via-screen. People were creators now. They were consumers. They were "netizens." They were leaders. They were centered in the story. They were celebrities-in-waiting.

The rise of the social web, just as *Time* intuited, has extended the possibilities of fame to more people than ever before. This is, in one way, both beneficial and overdue: Recaptured through social media's screens, people's talents have become direct-to-consumer products. Performance is becoming democratized along with fame itself. But exposure is a double-edged proposition. It creates ever new ways both to be seen and, as the revealing coinage goes, to feel seen. And as more people are experiencing celebrity, more are also bearing celebrity's risks. Fame elevates and belittles at the same time. Its new accessibility has meant that anyone might become a star. But anyone, too, might be dimmed.

The rise of mass media, the historian Warren Susman argues, changed the American character. It changed people's sense of what it meant to be a person. The shift, Susman argued in his 1984 book *Culture as History*, was exemplified by the rise of the radio. In the 1930s, in response to the availability and new ubiquity of radio waves, the advertising industry was beginning to mold itself around the power of the human voice. And so, Susman argued, were everyday people. Under the radio's influence, Americans would come to value what Susman called "personality": charm, likeability, the ability to put on a good show. "The social role demanded of all in the new culture of personality was that of a performer," he wrote. Soon, "every American was to become a performing self."

The insight was prescient. It was a suggestion, most obviously, of the long lineage of performance as an American cultural script. Today, though, what is perhaps most relevant about Susman's reading of the "performing self" is that it was such a departure from what had come before. In earlier eras, Susman suggested, Americans valued that collection of qualities that we might shorthand, today, as "character": honesty, courage, hard work. Those qualities were community-oriented. They reflected—and encouraged—individuals who would act for the good of the group. But the rise of mass media made "the group," itself, an ever more unwieldy proposition. The voices that were floating, invisibly, into people's homes and lives were not attached to communities. They were not attached to anything at all. Their charisma had no context. Being good at radio required no personal character. It required only that one *be* a character.

The essential pattern that Susman identified—community values fragmenting into individualism, under the influence of mass media—is, at this point, a fairly reliable trend. "Main character energy" may be something we observe in others; just as often, though, it is something we impose on ourselves. The scripts that shape our digital lives cast us, very often, into stardom. "Self-care," for example, arose within activist circles as a term of sustainability. The fights were long, the activists knew: The work of justice typically involves marathons rather than sprints. Self-care was recuperation that would re-energize activists in their work.

"Self-care" is now, primarily, a marketing term—a variation of that long-running L'Oreal line: "because you're worth it." In that sense, the term is one more among the many (see also: "gaslight," "trauma," "emotional labor," etc.) that have been defined away from their original nuance and scaled up into ad-friendly oblivion. But "self-care" has Susmanian dimensions, too: In its evolution, you can see a clear path. "Self-care," in its original form, was aimed at the good of the community. You took care of yourself specifically so that you would be able to take care of others. That value has been cleaved from the word. Now, for the most part, "self-care" speaks to consumers, rationalizing their desires. The out-of-budget handbag? The even-more-out-of-budget Botox? *Get them!* says the new self-care. Nothing is off-limits when the self being cared for is You.

Performativity, as a term and a broader concept, has followed a similar path. For Erving Goffman, the "performance of the self" was an element of an ensemble show. In the give-and-take of performance, in the sociological sense, people did not merely create new scripts; they also bolstered community. The contemporary theorist Judith Butler, translating those ideas to the work of gender performance, assumes that collective value, as well. Performance, in Butler's framework, is shared, but the sharing itself confers agency—and dignity—to the performers themselves. The performance is an ongoing interaction between the person and their social setting. Through the interaction, a person can construct, and constantly remake, their identity.

Gender, in that way, becomes "performative"—something, Butler writes, that is "put on, invariably, under constraint, daily and incessantly, with anxiety and pleasure." The performance is made for public consumption, yes. But it is not a spectacle. It offers no exemptions from reality. On the contrary: Performance, in some sense, is reality at its most essential. It is personal. It is political. It offers dialogue with the old scripts—and the freedom, always, to go off-book.

"Performative," as a mainstay of digital dialogue, all but reverses that sensibility. Summoned, typically, as an accusation—of attention-seeking, perhaps, or of insincerity, or of main character energy gone wrong—"performative" stifles community rather than fostering it. It assumes that the thing someone did was done for likes and clout. It leaves the accused, by and large, with no way to refute the claim. "I did it just to do it!" might be the truth; it might also be dismissed as one more element of the performance.

"Virtue signaling," a claim of performativity with a culture-war twist, works similarly. The term, used as an insult, effectively denies that virtue (known, in other contexts, as community-mindedness) might exist on its own terms. And it is especially effective as a barb—and especially cynical as rhetoric—because "virtue," itself, can seem suspect. ("The louder he talked of his honor," Ralph Waldo Emerson put it, "the faster we counted our spoons.") But "signaling" further twists the knife. Performativity is not an insult you might expect to catch on in a culture that so famously loves a good show. But even our accusations, now, follow the path of fragmented individualism. It's not the performances we're questioning. It's the performers. To perform, after all, is to ask to be looked at. It is to suggest that one is worthy of other people's gaze.

The Puritanism that prowls the back caverns of the culture tends to snap to attention when it detects a whiff of vanity. George Washington proved his worthiness for the presidency by declining to seek the office. Two centuries later, one of his fictional successors, *The West Wing*'s Josiah Bartlet, would do a version of the same: He ran for the White House only after his friend and future advisor recruited him for the role.

Fame, in American culture, has lived a split-screen existence. We seek it; we suspect it. We revere the famous; we resent those who seek the fame. We do not stop to wonder what distinguishes the two. But fame, overall, is a condition that is also an ongoing transaction. It is notoriously fickle. It giveth; it taketh away. And it is doing both, now, in the digital spaces we share. Fame's aspirations and aspersions are part of the internet's infrastructure. They make every action a performance, every setting a stage. They impose stardom on us, whether we seek it or not. They impose an audience, too—people who, whether they like us or follow us or friend us or rate us or mute us or cancel us, are giving over to the terms of the show. Digital spaces teach us to resent the spotlight; they make the spotlight inescapable. This is everyday celebrity. This is You, once *Time*'s person of the year, cast as a costar of the internet.

Time's announcement, in its physical form, followed the course that most print editions will. It gave way to planned obsolescence, thrown away, recycled, lost among the clutter. The few copies that remain tend to cede to their own form of entropy. The glue that held the foil-like strips loses its grip. The mirror that announced a bold new age of celebrity peels away from its anchor, edge by edge. The picture it reflects—you, the exciting new ingenue—becomes ever more warped. Wait long enough and the mirror falls away, leaving an empty space where a person had been.

"The celebrity is a person who is well-known for his well-knownness," the historian Daniel Boorstin argued in his 1962 classic *The Image: A Guide to Pseudo-Events in America*. Pseudo-events, as Boorstin described them, were cultural happenings that were manufactured for the needs of the camera. They would also come to be known as "media events." They were press conferences and press releases. They were Hollywood awards ceremonies. They were political conventions.

Pseudo-events, in Boorstin's framework, were happenings that were created specifically to become pieces of media. And here was his kicker: Celebrities were human pseudo-events. They were people whose duty, as public figures, was to be publicized. (Boorstin's formulations gave rise to the idea that people can be "famous for being famous.") Celebrities' direct work might involve acting or singing or dancing or entertaining. Their true role, Boorstin argued, was to serve as fodder—for conversation and gossip and, in particular, imagery.

Boorstin titled his book *The Image* for a reason. He published it two years before McLuhan would publish *Understanding Media*, and his book is rooted in similar insights about technology's cultural power. Celebrity, Boorstin argued, was a function of images. The celebrity of mid-century America, in his view, arose from what he called the Graphic Revolution: the series of technological advances that, between the nineteenth and twentieth centuries, transformed images from rarities into ubiquities. The revolution—which revolved around the advent of the camera but which included many other advances in industrialized image-making—did for pictures what the printing revolution had done for words. Its advances cheapened the production of images, which led to their proliferation in printed books, magazines, and other forms of media.

The Graphic Revolution, Boorstin argued, helped to create the modern advertising industry, transforming commercial messaging from a matter of straightforward need to one of perpetually unsatisfied longings. Images, in their own ways, revolved around the doubleness of "feeling seen." Whether photographs or lithographs or prints, images in one way offered straightforward representations of reality. But they also inevitably exaggerated reality. They offered depictions that were more dramatic and compelling and seductive than unvarnished could hope to be. Images are objects, in that way, that quickly become expectations. They create, and then expand, what Boorstin called "our mania for more greatness than there is in the world."

Celebrities operate similarly. They are facts of life who are also, when converted into images, larger than life. And that has always made for uncomfortable transactions. Celebrities were the original screen people. They have long lived in that liminal space between reality and transcendence, between fact and fiction. Celebrity itself, the critic Leo Braudy argued in *The Frenzy of Renown*, his 1986 study of fame, can double as a kind of tragedy. Fame is not something one attains, he suggests, but instead an always-elusive aspiration: something that, once found, always threatens to be lost again. Celebrity, in that way, is a constant struggle between pride and humility. It brings stars low both despite, and as a direct consequence of, their elevation. Surrender is the price of fame's power; to be "stellified," in Chaucer's term, is also to be dehumanized.

Stars, traditionally, have been figures of stage and screen, meaningfully separate from their audiences. The distance can amplify their allure; it can also subject them to ever stranger forms of stellification. They have been figures, relatedly, of a form of public ownership. A tabloid of his time referred to Charles Lindbergh, the world-famous pilot, as a "Grade A celebrity." He was, as such, "a public commodity, like gas or electric light."

The culture that has risen around celebrity in the intervening century has only ratified the bleak assumption that stars are, effectively, public utilities. They can serve, in that way, as cultural binding agents. The contemporary anthropologist Robin Dunbar describes gossip as a form of social exchange that can foster bonds within a social group—and celebrities, being familiar to so many, can be apt as subjects of communal conversation. Each tabloid story or TMZ tidbit or DeuxMoi update ratifies the idea that, in exchange for our devotion, our celebrities owe us access to them—to their images, their love lives, their breakups, their beauty routines, their political views, their homes, their children. The transaction has been fairly straightforward. We give them fame, and all that comes with it; in exchange, they give us themselves.

Celebrities constitute what the sociologist Francesco Alberoni called a "powerless elite": They exist to be talked about but have no sanctioned means of redressing slights of public opinion. They sign over their life rights, effectively, to the crowd. They live a version of the contemporary writer Ward Just's observation about ancient heroes: "Odysseus wept when he heard the poet sing of his great deeds abroad because, once sung, they were no longer his alone. They belonged to everyone who heard the song."

Celebrities, in other words, are omens. They have long led split-screen lives. Well before the rest of us would need to, they have been reckoning with what happens when people are mistaken for "people." Man's inhumanity to celebrity—the uncanny cruelty that so often befalls our stars—is on display in Jimmy Kimmel's occasional late-night series "Celebrities Read Mean Tweets About Themselves." It was there in the weeks before Kate Middleton, the Princess of Wales, announced her cancer diagnosis in 2024—as #WhereIsKate hurtled from observations (*hmm, she hasn't been seen in public for a while*) to demands (*show yourself, Kate*) to a series of ever-wilder conspiracy theories that claimed to account for her whereabouts (love triangles, spousal abuse, murder). The treatment was there, as well, in an edited-for-lols supercut titled "kim and kanye having a TOXIC relationship for 5 minutes straight."

The video, featuring scenes from the now-dissolved relationship between Kim Kardashian and Kanye West, was an artifact not only of their marriage but of our age. It was presented as a documentary; its genre, however, was dark comedy. It invited viewers to lol *for 5 minutes straight* at the *toxicity* of it all and then move on. The magazine *US Weekly* may insist that stars are "just like us." They are, celebrity culture has suggested, in every way but one: To become a celebrity is, on some level, to renounce your humanity.

That transaction has been mitigated, to an extent, by the fact that celebrity, in Leo Braudy's time as in Boorstin's, has also been an industry. People happily bore fame's costs because it came with so many benefits: wealth, power, public adoration. A celebrity is "a known individual who has become a marketable commodity," in the historian Simon Morgan's tart phrase. And the commodification, itself, has been a largely opt-in proposition.

Democratized celebrity only rarely comes with the traditional compensations. What it does tend to involve, though, are the human costs of fame. Everyday fame, also, can collide with another of celebrity's complications: Americans, in general, tend to hold extremely ambivalent ideas about fame. We revere fame. We suspect it. We value it as a commodity. We resent it as a broader goal. In the endless theater, performance is a constant expectation. As actors, we might offer ourselves up through our screens. As audiences, we might greet other people's offerings as clickbait, as fame-mongering, as bids for clout. We might complain that their performances have been insufficiently authentic.

"Performative" really does capture this tension. The word might now be a straightforward description. It might be an insult hurled at someone we perceive to be on the wrong end of authenticity. Both are ironic inversions of the term's sociological and political roots. Performative acts, for the theorist Judith Butler, are acts of agency and dignity: continual interactions between the person and their social setting that construct, and constantly remake, one's identity. Performance both creates and responds to communal scripts. It has no stars or audiences, but instead an ensemble cast.

This shared sense of performance, though, has been effectively lost in the encroachments of the endless theater. Performance, now, is generally assumed to be done for likes and clout. It is both an expectation and a cause of suspicion. "By being charming, you are lowering yourself," the journalist Janet Malcolm wrote in her final book, *Still Pictures*. "You are asking for something." Charm and fame are related propositions: They are bids for attention. They make that most enduringly controversial of requests: *Look at me*.

The tension of performance is captured, in an elliptical way, by a work of history written in the 1960s. *Anti-Intellectualism in American Life*, Richard Hofstadter's classic, considers one of Americans' deep-seated suspicions—not the suspicion of knowing things, but of *seeming* to know them and of admitting to know them. Anti-intellectualism solidified in American culture, the historian suggests, in part because Americans, from the earliest days, deemed knowledge itself to be self-centered. Knowledge was, in today's vernacular, inescapably *performative*. Learning, therefore, was performed rather than done for its own sake. Knowledge was something people showed off rather than something they simply had.

"The more learned and witty you bee, the more fit to act for Satan will you bee," the Puritan preacher John Cotton warned in 1642's *The Powring Out of the Seven Vials*. You might notice that "learned" and "witty" were, for him, effectively synonymous. Later: "I say bee not deceived by these pompes, empty shewes, and fair representations of goodly condition before the eyes of flesh and blood, bee not taken with the applause of these persons."

Learning, in Cotton's framing of things, is inherently pompous and showy and glib—because learning itself can only be done for the sake of the "shewe." His attitude was not uniform among the Puritans; nothing is ever so simple. But it was representative, Hofstadter suggests—an early form of what he called, in another book, the "common climate of American opinion."

The Puritans had carried across the ocean a protean version of the idea that "the personal is political." They deeply resented the British elites who had persecuted them, and they carried the spite with them when they landed on the new continent. They might have called themselves "people of the book," but they reliably excoriated bookishness. They were suspicious of the knowledge that came from books—not merely on religious grounds, but on the grounds that it was, in today's terms, a bad look.

On the whole, though, Hofstadter argues, the Puritan conflation of learning and "pompe" prevailed as the colonies became a country. It was there in the protopopulism of Jacksonian democracy. It was there in the literature of Horatio Alger and the many other authors who celebrated—and in many ways created—the trope of the self-starting striver.

Soon enough, the impulse that was rooted in American religious practice seeped into nineteenth-century class struggles. The by-the-bootstraps ideals that were prevalent in the popular culture of the time were practical matters, too: The middle class, and those seeking their entry into it, had no choice but to lift themselves up through hard work. (This was particularly so before the rise of public education—a system instituted, in part, as a redress of inequality.) Books were leisure. Books were upper-class. Books were, therefore, suspect. The Puritan ideals, you might say, collided with the Protestant ethic: As Americans valorized commercial success and the labor required to achieve it, they wove the old suspicions into the new economy. There were book smarts and there were street smarts—but only one would get you rich.

These framings are, in some ways, very familiar. (They are familiar, in part, because of Hofstadter.) In their interstices, though, Hofstadter's insights propel themselves into our moment. The "shewe" is there, always encroaching—always a reason to doubt. You can be smart in America, definitely. You can even be brilliant. You can be educated. You can know things. You can *want* to know things. But your knowledge, in the model Hofstadter provides, must always have an outlet. It must serve you, somehow. It must play a role in the endless performance.

"Sick of Being Perceived"

In 2010, the Library of Congress announced that its collection, going forward, would include messages that users had posted to the social network then known as Twitter. "Beginning with the first tweets of 2006 through 2010, and continuing with all public tweet text going forward," the library announced, public tweets—as opposed to the direct messages that are ostensibly private—would become a matter of national public record.

The aim of the move (an "exciting and groundbreaking acquisition," the library called it) was a noble one: to "preserve a record of knowledge and creativity for Congress and the American people." It also offered a new kind of evidence that Warren Susman's insights about the effect of the radio would translate, fairly seamlessly, to the world shaped by the social internet.

Twitter, at its creation, had been premised on intimacy, of a kind. The messages it hosted were limited to a mere 140 characters. (The platform was designed, at first, to translate the capabilities of SMS and texting to the web.) The service facilitated short—chirpy, twitter-y—updates and conversations. "Just setting up my twttr," the service's cofounder, Jack Dorsey, wrote, cheerily, in posting the world's first tweet. Early users' posts replicated the casualness: They offered updates about people's evening plans and their saddesksalads (and, soon enough—this being the early internet—their cats).

But the service scaled, and, in digital settings, changes in scale have a way of becoming changes in kind. The tweets became more complicated. They got, eventually, longer—through both extended character limits and the creation of threads and the mini-essays sometimes shorthanded as "tweetstorms." Users created—and then Twitter incorporated—the sharing function of the retweet. Journalists were early adopters of the platform, and they began using it for information-gathering purposes. In short order, the service that allowed people to text each other outside of SMS had become a rowdy revue: performers, crowds, lines that looked like exchanges but functioned, very often, as monologues.

When the Library of Congress announced that the

scripts of the show would be archived, then, the institution was capturing something about the theater overall. In one way, it was dignifying the tweeters as performers and scriptwriters. It was recognizing how much power there is in personal testimony, individually and at scale. It was celebrating every person who had sent a public tweet between the years 2006 and 2010 as a coauthor of the first rough draft of history. But it was also acknowledging—implicitly, at least—how thoroughly the terms of the show had shifted in that relatively short span of time. People thought they were writing casual updates; now, they were told that they had instead been writing history.

The sociologist Charles Horton Cooley, writing around the turn of the twentieth century, described the rise of the "looking glass self"—the tendency to perceive ourselves through other people's reactions to us. This moment has realized the concept literally. Never before have we seen ourselves so clearly and so incessantly. And never has self-consciousness been so literal a proposition.

Performance is so ingrained in American culture—as an aspiration, as an expectation, as a value—that it doubles as a script. Every day, we are being sold on our own stardom. We are being conditioned, both subtly and incredibly blatantly, into celebrity. A recent H&M ad campaign promised that the brand's clothes would make the customer "the main character of each day."

And main characters are made, for the most part, by

being looked at. After a pronounced decline in use since their heyday in the late 2000s, *Women's Health* reported, tanning salons are on the rise again—with a popularity fueled by young people who turned to influencers rather than doctors for their medical advice. (Or who listen to doctors but don't care about the risks: In a survey conducted by the American Academy of Dermatology in 2022, 28 percent of Gen Z respondents said that getting a tan was more important to them than protecting themselves against skin cancer.)

"Girls as young as 8 are turning up at dermatologists' offices with rashes, chemical burns and other allergic reactions to products not intended for children's sensitive skin," *The Independent* reported in 2024. They are doing what young people always will—seeking guidance for how to look, and how to be, in the world—but they are getting their advice, now, from marketers who act like friends.

Coming of age involves, typically, an ongoing struggle between experimentation and limitation. Boundaries are tested; boundaries are drawn. But young people are engaging in that process during a moment when many others—their peers and their mentors—are unsure where the boundaries should be drawn. In 2021, *The Wall Street Journal* published a report, based on leaked documents, detailing internal research that Facebook had conducted about Instagram. "Thirty-two percent of teen girls said that when they felt bad about their bodies, Instagram made them feel worse," the findings revealed. "Comparisons on Instagram can change how young women view and describe themselves."

These, on their own, were not striking conclusions. What was more revelatory was the follow-up finding: Many of the same young people who described Instagram as a source of degradation kept returning to the service anyway. The platform, for them, was less an optional hangout than an all-encompassing environment. Their friends were there. Their identities, in some sense, were there, too. Instagram, like so many similar platforms, makes louche use of the language of easy freedoms: *like, love, comment*. For these young users, though, it mimicked the terse transactions of celebrity. The price of seeing, on the platform, was being seen.

Looking and seeing are basic acts: so automatic, for those with the gift of sight, that they are barely choices at all. ("I can't unsee it!" we might wail, in mock agony.) But seeing, as a social proposition, is also a political act. The gaze is a mode of power. As the Covid pandemic officially ended—as "social distance" transformed from a public-health mandate into a broader kind of emergency—many people kept wearing masks. They did so not as defenses against other people's germs, but as defenses against other people's eyes. "I'm sick of being perceived."

Some reluctant performers, in response to the endless exposure, have been looking for ways to make celebrity optional again. Finstas—fake Instagrams—are images (selfies, typically) that are, by design, poorly lit and awkwardly angled and otherwise stridently unphotogenic. They are, for the most part, jokes: cheeky little anti-portraits, shared among small groups of friends. But they are also rebellions against Instagram and its endless expectations, hosted on the platform itself.

Not to be outdone, Facebook users have been practicing the art of "Realbooking": posting unflattering—or at least not self-aggrandizing—updates to their feeds. BeReal, meanwhile, brought the anti-selfie logic to a logical extreme: The app gives users a two-minute window each day, at varying times, to post pictures of whatever they're doing when the window arrives. "BeReal won't make you famous," the app says; "if you want to be an influencer you can stay on TikTok and Instagram." Not long after BeReal launched, the service met with the sincerest form of flattery: TikTok launched a competitor.

Self-Objectification

"To speak of reality becoming a spectacle is a breathtaking provincialism," the critic Susan Sontag wrote. "It universalizes the viewing habits of a small, educated population living in the rich part of the world where news has been converted into entertainment. . . . It assumes that everyone is a spectator. It suggests perversely—unseriously—that there is no real suffering in the world."

Sontag's point remains crucial and true, right down to the irony that the provincialism she described has been embraced by a global superpower. But the point might also be incomplete. Sontag published *Regarding the Pain of Others*, her final book and the one that contains these meditations, in 2003—a year before Mark Zuckerberg remade Harvard's student facebook in digital form, two years before YouTube gave people an easy way to transform themselves into pieces of content, and four years before Apple introduced the device that would make the internet and its disembodied people newly portable.

When Sontag wrote about the cynicism of the spectacle, she was thinking about war zones she had visited. She was thinking about empathy and sympathy and pain. But she was also thinking about television. The one-way screen divided people into groups of two: those who were watched and those who did the watching. To be watched was to be exploited. To be a watcher was to do the exploiting. The terms were blunt and left little recourse to the people recast as spectacles: Once your three dimensions are reduced to two, it's nearly impossible to become whole again.

But what becomes of those dynamics when the screens run in two directions? What happens to the hierarchies Sontag described in environments that dissolve the distinction between seeing and being seen?

"Even if I'm objectifying myself, I feel good about it,"

Kim Kardashian said in 2022. The reality star was responding to an interviewer's question about her stock in trade: the pictures Kardashian posts of herself, in varying states of dress and undress, across her social-media platforms. Kardashian, who has spent her career converting "famous for being famous" from an insult to an industry, has a knack for distilling cultural complexities down to single images and sound bites. *I feel good about it*—choice feminism, remade for the age of the influencer—was no exception.

On the web-connected screen, "objectifying myself" is not a contradiction. It is currency. People's willingness to remake themselves as spectacles is the basis of the social web's economy. Objectification's threats are offset, now, by an extremely web-friendly promise: *empowerment.* People, once be-screened—once remade as pieces of media—can speak for themselves, before potentially worldwide audiences. They can spark movements. They can join ones that are already there. They can broadcast themselves. They can explain themselves. They can assert themselves. They can maybe even humanize themselves. They can do all that because they have been objectified.

Kim Kardashian may feel good about those transactions. Those of us who have not profited so directly from them, however, might feel more ambivalent. We might consider the problem that the critic Walter Benjamin observed in the late nineteenth century, as industrialization brought the efficiencies of mass production to the creation of art.

Benjamin, serving as an eyewitness to Boorstin's Graphic Revolution, watched as images that had been scarce—and some that had been sacred—were churned out on factory floors. The machinery, he concluded, changed the product. It transformed what art would mean, to artists and everyone else. Even paintings have, in their way, three dimensions: swirls and slashes, high points and low, each texture a link to the absent artist. That the machines would fail to capture that was obvious enough; for Benjamin, though, the limitations were part of their power. The machines, unable to recreate the original, instead gave it competition. They gave art, as a category, a series of asterisks and caveats. The work of art would become "the work of art in the age of mechanical reproduction."

You might detect, in Benjamin's wording, a hint of menace. Art that was mass produced was art that was for the first time accessible to the masses. But it was also art that had been cleaved from the artist who created it. It was, in a new way, objectified. Art, so easy to romanticize as the product—and the proof—of the human soul, was both democratized and dehumanized.

If the tension feels familiar, that is because we are currently living our own version of it. The machines of the digital age, those tools of mass production, make things accessible and replicable and portable and disposable. The trouble is that, too often—and too easily—they do the same thing to us. On the screens, we are images. We are objects. We are pieces of media, disembodied. We are contending with the mechanical reproduction of the self.

Screens, as environments we inhabit, are utterly new. But they have been explored before. We might look for guidance in that age-old way: by looking to the stars. Some scholars have argued that the lofty people of the past—Greek gods, Catholic saints—were stars by other means. ("Celebrity," the word itself, is rooted in the Latin *celebritās*, a term connected with both celebration in general and the celebration of religious rituals in particular.) But celebrity as we tend to think of it today, that tangle of deification and indignity, has been made modern by way of a modern technology: the camera.

While earlier incarnations of celebrity were mediated through icons, paintings, monuments, and similarly brick-and-mortar forms, it was the photograph—newly lifelike and endlessly replicable—that allowed celebrity to adopt the distant form of intimacy that defines it today. Abraham Lincoln, for example, was beloved in his time not only for his leadership skills but also because his presidency coincided with the advent of photography: He was the first president whose face could be widely replicated, and thus widely recognized, by his constituents.

Cameras, too, helped to create the Hollywood celebrity of the twentieth century. In the early days of American cinema, the writer Anne Helen Petersen argues in her book *Scandals of Classic Hollywood: Sex, Deviance, and*

Drama from the Golden Age of American Cinema, film production relied on bulky cameras that required cinematographers to capture scenes' action from a distance. And the physical distance, as so often happens, amounted to a broader kind of remove. "Because viewers couldn't see the actor's face up close," Petersen writes, "it was difficult to develop the feelings of admiration or affection we associate with film stars."

When cameras improved, though, they transformed the relationship between the performer and the viewer. Lighter film equipment meant that camera operators could zoom inward toward the action—and the actors—on the set. This in turn allowed for a new kind of intimacy in the movies that resulted. Lenses, Petersen notes, could now emphasize actors' faces. They could, as a result, emphasize actors' personalities. As sound became integrated into the cinematic experience, actors' voices, too, became part of the show. Instead of lurching images, shot from afar, movie screens began to feature actors who looked ever more recognizable, ever more relatable, ever more human. The "star," as we know it, was born.

Celebrities live, then as now, in that liminal space between reality and transcendence, between fact and fiction. They function, for modern-day audiences, the way icons might have for people of the past—existing, as the film historian Jeanine Basinger put it in her book *The Star Machine*, "on some plane between ours and that of the gods."

This is also to say that they exist in the sometimes uncomfortable space between humanity and imagery. They are products, and embodiments, of a listless postmodernity.

Masked Singer

"Who, Who, Who, Who?"
One of this moment's era-defining TV shows is an exploration of those dynamics in the guise of campy reality. The Fox juggernaut *The Masked Singer*, a little bit *American Idol* and a little bit *CSI*, is a musical competition show in a fairly standard vein, with a twist: Its competitors, rather than hoping for stardom, are people who have already attained it. And the celebrities, rather than performing as themselves, offer their renditions of "Fight Song," "Stay with Me," and the like while disguised in elaborate costumes.

Each week, a collection of cosplaying stars—a unicorn, maybe, or an alien or a pineapple—appear before the show's cheering studio audience. And each week, the lowest-performing member of the cast, as determined by the votes of the audience and the show's panel of judges, leaves the show. The climax of each episode is the ritual that accompanies each departure: As The Who's "Who Are You?" blares ("WHO, who, WHO, who?"), and as the judges and crowd chant—"Take! It! Off! Take! It! Off!"—

the masked singers unmask themselves. The famous people made anonymous regain their fame once more.

The Masked Singer is postmodern in its sensibility and *The Hunger Games* in its aesthetic: It is artifice, all the way down—and, then, back up again. Critics have maligned it as "a deranged reality TV fever dream" and a source of "unnameable dread." Perhaps in reply, the season of the show that aired in late 2024—its twelfth—debuted as one of its "characters" a blob of green slime that wore a sassy bow tie and an expression of permanent shock. Its name was, simply, Goo.

But *The Masked Singer* has achieved the status that it has—it is not only a fixture on Fox's prime-time schedule, but also the source of multiple Reddit threads and *People* articles and podcast segments and "What We Know So Far" summaries on *E! News*—because its derangement, in the end, is a feint. The show's true genre is less "singing competition" than it is "mystery"—a whodunnit in which celebrity itself is the victim and the culprit. "Whooooooo are you?" is not merely the show's climactic reveal but also its primary selling point. After each masked singer performs, the show's host, Nick Cannon, and its panel of judges discuss the performances—but the bulk of their assessments are forensic rather than critical: They, too, are trying to guess the identities of the celebrities behind the (full-body) masks.

The show invites its audiences—Cannon sometimes calls them "detectives"—to do the same. It provides, as clues, elaborate biographical packages that, through teasingly elliptical wording, share obscure details of contestants' lives. It treats its performers not just as celebrities, but also as sources of shared trivia in the Robin Dunbar vein. Did you know that Terry Bradshaw raises horses? Or that La Toya Jackson has had pet snakes—and that she named them Adam and Eve? Or that *NSYNC's Joey Fatone, in addition to his membership in the boy band, has also run a hot dog empire? *The Masked Singer* thinks you might. It knows that celebrity is not merely an aspiration but also a set of banal facts—many of them gleaned from cultural osmosis. In that, the show premised on disguising celebrities ends up revealing their omnipresence.

The Masked Singer premiered in 2019 during a cultural moment that brought a new anxiety to the question of what celebrities deserve. Many products of the time—documentaries, essays, books, podcasts, more—were reassessing the treatment that celebrities (particularly women) had received in earlier eras (particularly the 1990s). Britney Spears, Pamela Anderson, Anita Hill, Princess Diana, Janet Jackson, Monica Lewinsky, Lorena Bobbitt, and many other famous-and/or-infamous women were reconsidered in light of the #MeToo movement. So were Michael Jackson and Bill Cosby and Mike Tyson. The '90s' "trial of the century" took on a renewed life in this one, as filmmakers and journalists revisited the murders of Nicole Brown Simpson and Ronald Goldman and the subsequent trial of O. J. Simpson.

The reprisals often doubled as indictments—of sexism, of racism, of a media apparatus driven by the assumption that exploitation amounted to good ratings. (The heart of the Simpson trial was a gruesome murder that left kids without a mother and families without their loved ones. It was also, as the TV producer Don Hewitt had declared at the time, "TV's longest running entertainment special.") But many of the projects doubled, as well, as acts of atonement. Here were celebrities who had been mistreated, misunderstood, misappropriated. And here were belated admissions of the errors. The projects, in that way, also suggested self-congratulation: *We've come so far since then*, they promised. *We know so much better now.*

The celebrations, however, were premature. An early finale episode of *The Masked Singer* featured a character known as The Monster. The creature—he was played, it would soon be revealed, by the rapper T-Pain—had the appearance of a sports mascot gone feral: mint-green and furry and conically Minionlike in form, featuring a single eye, a pair of teeth, and three fingers that wiggled on each paw. After performing a showstopping rendition of Sam Smith's "Stay with Me," The Monster waddled over for a chat with Nick Cannon. "Deep down," the host asked, "*who are you?*" The Monster, his speaking voice disguised in a decidedly unmonstrous squeak, replied, "I'm a father, a husband, a son, a brother—and more than anything, I'm a *person*."

Lines like that are a mainstay on *The Masked Singer*.

Through them, this self-consciously effusive paean to stardom gets punctured with Braudian agony. "I'm a *person*" is a declaration that is also an argument. When it needs to be said at all, though, the fight has already been lost.

The Truman Show

The web is affecting public spaces in roughly the same way that it is affecting so much else: It is changing those spaces from a fact into a choice. Public spaces now have competition from all the entertainments and distractions and possibilities that thrum within our portable screens. Screens complicate the "public" element of public spaces. And the complication is exacerbated by the fact that the screens can record the interactions.

And the people we encounter in those spaces become options, too. You can interact with them. You can ignore them. You can be unsure what the courteous path might be, anymore. Is it more polite to engage with people? Or is the polite thing, indeed, to ignore them?

Minidramas like that take place every day, now, in cities and towns, in stores and restaurants and airports and pretty much any public space. They arise because there's still so little consensus about screens' proper place in those spaces—even as the screens themselves proliferate. Soon enough, it stands to reason that your efforts to write about inescapable screens would be interrupted by one more inescapable screen.

The law professors Michael Heller and James Salzman begin their 2021 book *Mine! How the Hidden Rules of Ownership Control Our Lives* with a meditation about airplane seats. "Is it rude to recline your seat?" has been a perennial question, particularly once the web gave people a forum to debate it at length. One reason the debate never ends, the authors suggest, is that the few precious inches of space behind each seat are effectively co-owned by the two people who share it. The occupant of the forward seat can thus claim a right to recline; the occupant of the seat behind can, with equal validity, claim a right not to be reclined into. There is no good way to litigate this, Heller and Salzman suggest, because the fine print of the ticket will typically go out of its way to keep the terms vague.

Public spaces writ large have long been similarly ownership-ambivalent. But shared spaces are greatly complicated by the fact of our screens. And those terms are complicated further still when the screens can turn the people we encounter in public into pieces of media. Anyone, today, might be stellified.

Democratized celebrity is one more step in the Graphic Revolution: It arises from advances in photography. Portable cameras; ubiquitous cameras; front-facing cameras; the ease with which a person can be transformed into a shareable image, by themselves or someone else—these are the technological changes that are transforming the nature of fame. And they are encroaching not just on the world of the screen, but into the physical world, as well.

In 2012, in a paper published in the journal *Cognitive Neuropsychiatry*, Joel and Ian Gold discussed what they called the "*Truman Show* delusion." Their coinage now goes by other terms (among them "Truman syndrome" and "main character syndrome") and is often invoked as an accusation of someone else's self-centeredness—"main character energy," gone wrong. But those senses all but reverse what the brothers—the former a psychiatrist, the latter a philosopher—were getting at when they coined the term. The "*Truman Show* delusion," in their sense, was a fear of being watched rather than a demand of it. It arose from an anxiety Joel had observed among several of his patients: the belief that they were under constant surveillance, their movements recorded by unseen cameras. The brothers named the delusion after *The Truman Show*, the dark comedy about reality TV gone awry, because the fear of being endlessly watched tended to be accompanied by another: the fear that their lives were being endlessly broadcast for other people's amusement.

3
The Sets

The Stanford prison experiment ended the way it did partly because of human nature and partly because of *Cool Hand Luke*. The widely taught and cited study, conducted by the psychology professor Philip Zimbardo in the summer of 1971, recruited twenty-four young men to play the roles of inmates or guards within a simulated prison: a setup intended to test the limits, or the limitlessness, of human cruelty. In the lore, the experiment proved its thesis so thoroughly—the guards behaved with such brutality toward the inmates—that it had to be ended early on humanitarian grounds. In truth, several of the guards now say, the only thing the experiment proved was their acting ability. "I took it as a kind of an improv exercise," one participant later admitted. Said another: "I faked it."

The Zimbardo experiment's initial findings—an illustration, Zimbardo would later claim, of "how good people turn evil"—would help to make the study one of psychology's most famous and infamous. But no experimental control could counteract the force that exerted itself within that manufactured prison: the assumption that every setting is best understood as a set. "The first night was boring," one participant, a guard, later said. To compensate, he calibrated his performance. He channeled the sadistic warden from *Cool Hand Luke*, the popular prison drama, yelling and threatening and applying method acting to the scientific method. Fellow guards, they say, did versions of the same. Zimbardo's study, for years, seemed to offer a quantification of man's inhumanity to man. What it really provided, in retrospect, was evidence of man's deference to the show.

Under screens' regime, the physical world is remade, day by day, into a set for endless performance. There are extreme examples (the influencer who staged a photo shoot before the open casket at her father's funeral; the tourists who vamp their way through Holocaust memorials). But the breadth of it all, at this point, is also the stuff of cliché. Many weddings now have, effectively, stage managers. The productions that result—some of them financed with the help of high-interest "wedding loans" that turn a daylong party into ongoing debt—can operate as split-screen affairs: celebrations of love and hours-long studio shoots staged for the cameras. The gender-reveal party—an adjunct to a baby shower—has, in recent years, gone from not a thing at all to a thing that, via fireworks of pink or blue, produces regular environmental emergencies. (The guests of both types of parties might be "curated" for the occasions.) Marriage proposals, in the social-media age, are evolving from questions into acts of content-friendly stagecraft. "Promposals" are allowing teenagers to participate, as well.

Soon enough, "doing it for the 'gram," the old ethic of Instagram, becomes, simply, living for the 'gram. Screens' alchemies come for the world at large, transforming it from a place that exists on its own terms—a place that is communally shared—into a backdrop for the endless shoot. Nature transforms from a place to be experienced into a place to be occupied. Places themselves, reimagined as sets, lose their historical context and their human meaning. People come to expect life itself to adhere to the demands of the show: to be fascinating but convenient; to be beautiful but sanitized; to be dramatic but not *too* dramatic. We expect it to entertain us without implicating us. We expect it to delight us without troubling us. Which is to say: We expect life to act like the web.

In the age of the relentless lens, Instagram is a constant architect. Buildings, many created with creators in mind, serve as backdrops for selfies and other photographs. Neon signs, often with chirpy messages (GOOD VIBES ONLY, etc.), litter the walls of restaurants, bars, gyms, beauty salons. In Nashville during peak tourist season, so many people vie to experience the artist Kelsey Montague's "What Lifts You" mural—a massive pair of wings, with a selfie-friendly space between them—that a line can form. The wait, Anne Helen Petersen reported, can take ninety minutes.

A hotel room I stayed in recently fully committed itself to the idea that a physical space could be both a set and a director at the same time. MAKE IT A GREAT DAY was framed as art on one of its walls. Next to that was another frame: BE KIND. Even the bathroom mirror beckoned like a stage manager who wants to keep the talent ready for the performance to come: HELLO GORGEOUS, it said.

Other people are not merely the audiences for the shows; they are part of the scenery, as well. In 2024, a thread went viral on X. A poster on the site had complained about the customer-service experience at the grocery chain Trader Joe's. The stores' workers, particularly those who help customers with the checkout process, are famously chatty: Small talk is part of the Trader Joe's job description part of the service the chain tries to offer its customers. For this particular customer, though, the small talk was a customer-service failure: Why should buying groceries involve chatting with strangers?

Trader Joe's has, so far, bucked the trend that other grocery chains have been embracing: the self-checkout kiosks that allow customers to buy their goods with minimal—

typically zero—human interaction. But the expansion of self-checkout has meant that, particularly for younger people who have no other frame of reference, the interaction itself can seem like a violation of the norm. As one X user put it, in response to the thread: "I really feel like the internet has raised a generation of people who expect the entire rest of the world to be NPCs."

NPCs—"nonplayer characters," in the parlance of the video game—are characters who are, quite literally, part of the scenery. Their purpose in the game is to give the illusion of interpersonal interaction without the inconvenience that can come when people actually interact. Derek Guy, the fashion expert otherwise known as "the menswear guy," weighed in on the conversation, as well. "This is something that sales associates in clothing stores have to deal with a lot," he wrote on X. "As more ppl shop online, many are unaccustomed to the high-touch service that once marked retail. Some are even hostile to the idea. So associates have to be careful of who they approach and how."

If screens can alienate us from our bodies, sets can alienate us from one another. Self-service is a principle of engagement as well as design; it has arisen from the assumption that customer service is at its best when the service itself isn't human. The assumption, I have to admit, can be accurate. Given a choice between a mechanized kiosk and a standard checkout lane, I will often opt for the one that lets me scan and bag and go. The ease of it can be convenient.

It can avoid the risk that I might be judged for a haul that consists of one head of organic kale and three pints of Ben & Jerry's Peanut Butter Cup.

At scale, though, my personal choice of lane—combined with everyone else's choices—becomes an imposition. It can lead to workers losing their jobs. It can lead to companies finding ever more canny ways to automate people out of their livelihoods and to automate the humanity out of commerce. It can lead to a world that, by design, discourages human interaction, writing it off as an inconvenience to be overcome rather than the whole point.

Main character energy, invoked as a principle of set design, might seem to elevate us: the world itself, offering endless assurances of our centrality to the show. Instead, all too often, it isolates. It denigrates. The stage, as it takes ever more cues from the screen, divides us into ever more roles. We might be stars or extras; we might also be cast as supporting players in other people's shows.

OnlyFans turns users into directors (often, though not always, of pornography). GrubHub, Seamless, UberEats, and other food-delivery services replicate the contact-minimal "magic" of the self-service aisle at a location of the customer's choosing. Amazon products are delivered by mostly unseen laborers. Cameo and similar message-sending services fully invert the traditional dynamics between the star and the fan: Their users can pay celebrities to film short videos that follow scripts that the users have written. Actors, podcasters, influencers, and politicians have used such services as alternate revenue streams, some quite successfully. (The former vice presidential candidate Sarah Palin, *The Washington Post* reported in 2022, earned more than two hundred thousand dollars through the platform the year prior.)

Starbucks baristas will sometimes engage in what has become a rite of passage: They'll share the absurd orders that have been made through the company's app. The mobile menu allows customers to customize their drinks in a nearly infinite variety of ways; the concoctions that sometimes result from the merger of entitlement and infinitude involve roughly six kinds of syrup, four kinds of milk, and in general an unholy fusion of sugar, caffeine, and dairy. Sometimes, the customers will specify that they want three pumps of peppermint syrup, five pumps of toffee nut syrup, and four pumps of mocha. Sometimes, they will also specify that they want their beverage topped with only half a serving of whip, but double a serving of mocha chip sprinkles.

The drinks may be fun for the people who order them; they are also insults to the people who will bear the weight of all the whimsy. Working a Starbucks counter is a hard and hectic job; very few customers, ordering in person, would have the temerity to request a drink that will take someone 35.5 steps to make. But the app mystifies the labor so completely that the laborers themselves become subject to the forces of unreality. The baristas are unseen by the app-using customer, and the absence renders them, effectively, unreal. After all: It is the customer, not the worker, who is the main character of every day. And if there's one thing everyone knows about main characters, it's that all the action—and the rest of the cast—revolves around them.

The social psychologist Stanley Milgram might be best known for the controversial experiments he conducted in the 1960s: tests designed to measure humans' deference to authority. (The tests, sometimes shorthanded simply as "the shock experiments," found that majorities of participants were willing to administer painful electric shocks to others when instructed to do so by perceived authority figures. Only later was it revealed that the shocks in question were simulated.) But Milgram was interested in social bonds as well as breakages. In the early 1970s, he and a team of students conducted research into what he termed the "familiar stranger": the people you might recognize as repeated figures in your daily life but with whom you don't directly interact. Familiar strangers can be the people you see on your commute or those you might encounter at a lunch spot or in the supermarket.

Familiar strangers are defined by distance—once you interact with them, they cease to be true strangers—but they are also defined, in Milgram's view, by a kind of silent solidarity. The strangers are part of life's scenery, yes, but they are more than NPCs. When Milgram and his students interviewed commuters about their relationships with familiar strangers, the commuters commonly mentioned that they imagined the lives the strangers might conduct. The commuter, Milgram suggested, conducted with the stranger a "fantasy relationship that may never eventuate in action."

In one way, Milgram was describing parasocial relationships: the unreciprocated bonds that might connect fans, say, and celebrities. But these relationships were typically mutual. And when familiar strangers encountered each other in contexts different from their typical ones, Milgram found—if, say, they ran into someone they recognized from their weekday commute at a movie theater on a Sunday—they were more likely to interact directly: "Hey, don't I know you from . . . ?" The "familiar," in that new setting, would win out over the "stranger."

But the architectures of everyday life, now, are eroding the possibilities even of those meager types of interaction. And those architectures are matters of more than buildings alone. Earphones—whether bulky or tiny, whether "noise-canceling" or standard-issue—are walls in wearable form. They can serve as armor, preventing against the microviolations that can come with existing in public settings. They might discourage people from catcalling women on the street. They might allow people to focus on their podcasts or audiobooks, spared the blarings of car horns or the rumble of construction work.

But headphones can be barriers, too. They can double as Do Not Disturb signs, capable of transforming a personal thing—your desire to listen to a podcast on your commute, say—into a social one: the foreclosure of even minimal interaction with the people around you. As with so many features of the digital age, the costs and the benefits of headphones can be difficult to untangle. As with so many features of this moment, as well, the devices tend to resist nuance. "I'm listening to a podcast but feel free to interrupt if you need directions or just want to chat or something!" is not a message headphones are great at communicating. (Nor is the corollary: "But feel absolutely unfree to interrupt with catcalls, obviously!")

Those are, in many ways, minor tensions—questions of everyday etiquette, rising in reaction to the changing stage. But they are consequential, unavoidably so, because the "noise" headphones cancel so often includes other people. Matters of etiquette can be matters of ethics in disguise, their uncertainties distilling those most basic questions of all: Who matters? Who doesn't? In the hybrid world, we are so unsure how to regard one another that even courtesy becomes a question. Is it more polite to engage with people? Or is the polite thing, indeed, to ignore them?

The Whole World (Could Be) Watching

In 1791, the philosopher Jeremy Bentham published his design for the Panopticon—a circular prison whose capacities for ceaselessly mutualized surveillance, he thought, would regulate prisoners' behavior. The prisoners would live in cells along the perimeter of the building, Bentham proposed, with their guards stationed in an inner circle. The guards, in this arrangement, would be able to see the prisoners, without the prisoners seeing the guards. Well before HGTV came along, Bentham had intuited both the promise and the flaw of the open-concept space: Nothing could be hidden from view. The architecture itself would regulate the prisoners' behavior, Bentham thought, because the prisoners would never know whether or when they were actually under the guards' surveillance.

The Panopticon was later examined by the twentieth-century theorist Michel Foucault, invoked as a warning about the skewed power dynamics between the individual and the state. Today, it is commonly invoked as a metaphor: a way to encapsulate the sense that we are subject to creeping spectatorship—in ways we are both aware of and, like Bentham's prisoners, not privy to at all.

To be out in public has always implied the risk of exposure to others, as well as the gift of it; "see and be seen" is a cliché for a reason. The inherent vulnerability of publicness, however, now includes the possibility of seeing and being broadcast. In a 1947 article for the *St. Louis Post-Dispatch*, the critic John Crosby announced the arrival of a new era of American entertainment: the age of "the involuntary amateur." The occasion for the announcement was the popularity of *The Candid Microphone*, the radio show that aired on ABC on Sunday evenings. The program, conceived and hosted and generally evil-geniused by the entrepreneur Allen Funt, was an audio-only precursor to *Candid Camera*, which Funt would bring to TV in 1948.

Like its televised successor, *The Candid Microphone* was notable for both its inanity and its historical significance. The show subjected unsuspecting people to various pranks that Funt had dreamed up for the occasion, their reactions captured by hidden microphones. Funt had gotten the "secret recording" idea during his World War II military service: As a morale-building exercise at an Oklahoma Army base, he created a "gripe booth" in which soldiers could vent their grievances. Recording the sessions, but finding that many of the appointed gripers would become nervous knowing that they were being recorded, he found an elegant solution: He hid the recording devices.

The Candid Microphone was a zany comedy show: a primordial *Punk'd*. It was a pioneer in part because of its innovative use of technology. (Microphones light enough to be made portable and cameras small enough to be made hidable were cutting-edge capabilities in the late 1940s.) But the series, with evocative prescience, also straddled— and in some ways invented—the line that reality TV draws between elevating people and mocking them. At the end of each episode of the radio show, Funt invited listeners to write in with specific names "if there is someone you'd like to hear us catch off-guard." He would later describe *Candid Camera* as a project of quasi-sociology: a show dedicated to "catching people in the act of being themselves."

Candid Camera proved so durable as a reality TV concept that the show was regularly revived well into the twenty-first century. (The most recent reboot aired in 2014 and was cohosted by Peter Funt, Allen's son, and the actor Mayim Bialik.) But the series' sociological bent has proven equally timeless. Both *The Candid Microphone* and *Candid Camera* captured the ease with which recording devices, made portable and invisible, could transform public existence into a potential threat. The devices could remake the person on the street as Crosby's "involuntary amateur," their confusions and frustrations treated as fodder for national hilarity.

Participation in the shows was technically an opt-in endeavor; releases, then as now, were required before people's likenesses could be aired. But as the critic Emily Nussbaum points out in her book *Cue the Sun! The Invention of Reality TV*, the "choice" given to recordees wasn't much of a choice at all. People were sometimes cajoled and harangued into participation: The show had invested in them simply by recording them. And "be a good sport" can be a pretty powerful argument. The "age of the involuntary amateur" had arrived—not just as a TV show, but as a fact of life. As John Crosby put it, "the possibilities are limitless; the prospect is horrifying."

Today's cameras, rarely asking permission and always hiding in plain sight, have made that prospect banal. Little wonder, then, that many people are simply opting out of public settings. An entire genre of meme has arisen from the relief people claim to feel when plans are canceled, giving them an excuse to stay home. ("Me when my plans get canceled: [gif of a kid dancing/a dog wearing sunglasses/etc.]") "Cocooning" and "bed-rotting" have become something of a performance art on social media, as people describe the appeal of staying home rather than going out into the world. Hanging out, that low-stakes and time-honored form of socializing, has recently become so endangered that, in 2023, the professor and critic Sheila Liming offered a book-length defense of it.

In early 2025, my *Atlantic* colleague Ellen Cushing declared a national "party deficit." One of the many bleak statistics she cited to make the case: On an average weekend or holiday in 2023, only 4.1 percent of Americans attended or hosted a social event. This was a 35-percent decrease from stats gathered in 2004.

There are many reasons for the decline (one being, as Ellen noted, a simple disconnect between people's desire to attend parties—still pretty large!—and their desire to organize them: much smaller!). But the memes suggest that the performance mandate might explain at least some of the shift. Parties, like publicness in general, come with new vulnerabilities. They carry the risk of "being perceived."

They carry the risk of being filmed and posted—of being turned into someone else's clickbait.

But they also carry the risk of something that sits ever more uncomfortably with the demands of the endless stage: awkwardness. Performance tends to imply perfectionism. Its elements are rehearsed in advance so that, when they are finally presented to an audience, they will be shed of mistakes. When performance becomes a mandate, those standards can leak into any social interaction. The desire for a set that is under total control—in which every line sparkles, in which no awkward pauses are allowed, in which misspeaking and misunderstandings are violations of the show—is a rational response to parties gone panoptic. But it is one more way that our connected screens can keep us at a distance. When the stage never ends, the stage management doesn't, either.

In the months after Covid vaccines were obviating "social distance" as an epidemiological mandate, a new term began spreading on social media: FODA, short for "fear of dating again." The hesitance many people expressed about returning to dating (this would come to be known as, yep, "hesidating") was in some ways related to Covid itself. Exposure to people in person could still mean exposure to the virus. But the fear also came from anticipatory awkwardness—the sense that, over many months of social isolation, they had lost their capacity to conduct conversations that would be suitably screen-friendly.

The fear, whether it involves dating or broader interactions, can feed on itself. If you don't interact with people regularly, each interaction can bring more pressure to "perform" well. And the less you're around other people, the less patient you might be of the foibles that can compromise the performance (whether they're your own or someone else's). The cycle would be elegant, were it not for the human consequences. Jeremy Bentham, imagining a prison where inmates guarded themselves, wondered whether surveillance could be remade as a setting. It would take centuries, but he has his answer.

"There was of course no way of knowing whether you were being watched at any given moment," George Orwell wrote in *1984*, of the "tele-screens" that operate, in the novel's dystopia, as one-way Panopticons. Every moment, under Big Brother's watch, is saturated in surveillance. But tele-screens are not merely tools of out-in-the-open espionage. They are also TV sets. They air dramas and comedies and interactive exercise classes. The purpose of all this—Big Brother, with a content arm—is not merely to distract the subjects from the fact of their surveillance. It is also to make the surveillance so natural—such a plain feature of everyday life—that it will seem less like a violation and more like a sacrifice made in the name of all the fun. In 2022, the New York governor Kathy Hochul announced that New York City would be adding security cameras to the city's public transportation system. "You think Big

Brother is watching you on the subway?" Hochul said of the decision. "You're absolutely right."

The move is framed, as such moves often will be, as a security measure. It might be more accurately called security theater, though, since it is, at the very least, redundant. Any person who carries a smartphone is a person who has become, in a sense, a roving camera. That can lead to immense good: People acting as ad hoc documentarians have, for one thing, made the breadth of police brutality in this country newly legible. It can also lead, at times, to less straightforwardly just realignments of power. In a public place, one might always be ensnared in someone else's lens. One might be taken out of context. One might be edited as a hero or a villain or a joke. One might be made to dance for the camera. In an ad aired a few years ago, the NFL and its never-ending PR campaign tackled the fourth wall. "Every Sunday you watch us," the spot intoned. "But we want you to know: We're watching you, too."

The reminder was hardly necessary. Even at home—that allegedly most private of spaces—people are ever more onstage. Services like Siri and Alexa, in the name of being endlessly helpful, are also endlessly listening. Venmo, the money-transfer app, makes many of its financial transactions public by default. Ride-share services, as they "disrupted transportation," ushered in a quieter innovation: the mutualized ratings system that made both the driver and the passenger subject to each others' stars. The shift (slightly) rebalances the dynamics between the service provider and the customer, fighting back against the glib entitlements of "the customer is always right." But it also means that both parties, in effect, have an audience.

The sense of an ambient audience is there, as well, in the ellipses that pop up in chat apps when your cochatter is typing. It's there in the knowledge that your own typing renders similarly. Great art has been made about the particular agony that can descend when the dots stop and the chats end (see also: text-message read receipts and the ache of being "left on read"). Even in banal circumstances, though, the dots are text that come with inevitable subtext. Typing, now—the dots that flow; the dots that end—has become an observable act. It has become, as such, an analyzable one. Dot-dot-dots that start and stop, pausing and returning? The typer must really have taken their time with the message, considering it, editing it, investing in it. The chat-room message announcing that "several people are typing"? Something big must have happened, for all those fingers to be fluttering.

For many workers, too, job performance can now involve smaller acts of performance—for bosses and coworkers, sometimes, and for the amorphous authority of the algorithm. Google Docs and their adjuncts allow word processing, data processing, and more to play out on several screens rather than one. Even as they allow for new kinds of collaboration and efficiency, they can also mean that editing and other work—tasks that revolve around the idea that the first draft might not be the final one—gain an unrequested audience. Google Docs in that way are marvels of group creativity that might also have invented a new genre of performance anxiety.

More nefariously, spyware installed on many corporate computers might track workers' keystrokes, monitoring their movements (and thus, ostensibly, their productivity), often without their knowledge. Access to corporate offices might now involve smartphone apps: a convenience for workers (never again a forgotten fob!) that also allows their comings and goings to be monitored with unprecedented ease. The surveillance can be even more extensive for employees whose jobs involve more than information work alone.

"You feel like you're in prison," Wendy Taylor, a packer at Amazon's STL8 fulfillment center in Saint Peters, Missouri, told *The Guardian* in a 2024 interview. Taylor was one of several warehouse employees who had joined to file an unfair labor practice charge against Amazon. (The case remains, as of this writing, under consideration with the National Labor Relations Board.) "They know every move you make," she said, "when you're working, when you're not working. They surveil you with their cameras. Managers surveil you with their laptops because they can pull up your profile and a bar changes a certain color when you're not active."

The workers' claims suggest "feeling seen" in its most destructive form: a violation that comes not simply from being observed, but from, allegedly, being tracked. Social scientists talk about the web's creation of "ambient awareness": the generalized sense of other people's movements that can accompany life online. Such awareness, though, can have a corollary: To be aware of other people's digital movements is to be aware, as well, that your movements are similarly monitored. On the web, you are an element of other people's scenery. You are part of other people's vibe. You might be consumed, directly, as a piece of disposable content. You will be consumed, in ways both vague and inevitable, as an ambient data point: one dot among many, helping to form someone else's impressionistic picture.

Algorithmic Surveillance

And the "someone else" need not be human. Online, nearly every move we make, whatever our reason for making it, has a second life as a script: a piece of character development that will inform the algorithms' picture of who we are, as the web's residents and its consumers. I recently clicked, mindlessly, on an article that mentioned Prince Harry; the algorithm mistook this as an indication that my ideal version of the web would teem with Harry-related content (Harry himself! The royal family in general! Meghan! Polo!). For months afterward, thanks to the algorithms that double as set designers, I navigated a Harry-fied landscape.

I was, in the eyes of the algorithm, that classic character of urban life: the familiar stranger. But there was no mutuality to our relationship. Instead, like the city dwellers that

Stanley Milgram analyzed in his experiments, the algorithm was imagining who I was from afar, based on incomplete data. It was conducting a "fantasy relationship" with me. I could not reciprocate. I could not explain myself. In the digital city, "see and be seen" is a profoundly one-sided proposition.

In the early twentieth century, the sociologist Georg Simmel turned the post-industrial city into a site of ongoing investigation. Urban environments in late-nineteenth-century Europe were teeming and hectic and dirty and loud. Many had experienced rapid expansions of their populations as new employment opportunities lured people away from their rural lives. Many, consequently, had more people than they had space to contain them—and that meant, in turn, that many people were exposed to one another in ways they hadn't been before. They were contending with, as Simmel put it in his 1903 essay "The Metropolis and Mental Life," "a faint sense of tension and vague longing, a secret restlessness"—a "helpless urgency" that "originates in the bustle and excitement of modern life." In the process, they were inventing new ways to see and be seen.

One of Simmel's core observations neatly foregrounded McLuhan's theories: The strategies people developed to negotiate with the city were shaped, very often, by the structures of the city itself. The thrum of urban life—that endless supply of new scenery and new novelties—encouraged some people, Simmel observed, to seek "momentary satisfaction in ever-new stimulations, sensations and external activities." The frenzy encouraged others to retreat, even when they were in public. They tried to insulate themselves through clothing and behavior. They tried to recreate, as they moved among the crowds, the protective walls of the buildings.

City life, Simmel suggested, also gave rise to new categories of public personhood. One of them was "the stranger"—a figure rarely encountered in rural settings and omnipresent in urban ones. In Simmel's framework, the stranger is not merely someone unknown. Instead, the stranger is part of the cityscape, and an embodiment of the intimate distances of metropolitan life. Strangers are figures of doubleness and contradiction. They are, in physical terms, near enough to be observed but far enough to defy scrutiny. They are not in one's social circle, but potentially connected to it through unknown modes. They are part of the crowd and separate from it.

Strangers as Simmel describes them are, ironically, familiar. They are, in their intimacy and in their distance, the defining figures of the internet. But Simmel's definition is complicated by the fact that our environments, urban or otherwise, are web-connected. The neither-both qualities of the stranger as he defined it included an abiding sense of mutuality: Just as you could see them, from your intimate distance, they could see you.

Today, though, the people you encounter in public carry with them the power to expose you to millions of other strangers. The stranger, today, might take a creep-shot of you and upload it to the internet. The stranger might record you, from two tables away, as you're on your awkward date. The stranger might treat you as set dressing in their particular show. Or: You might appoint yourself as someone else's videographer. You might be the stranger who converts another person into a piece of content—hoping that, through them, you will go viral.

The web means that everyone, in some sense, lives in a city now. We are charged, every day, with doing a version of what the algorithms try to do: We try to make sense of people from a distance. We navigate other people, as sharers of public spaces and as features of our daily scenery. We try to make them legible to us, even from afar. Our efforts might lead to connection. They might also lead to alienation. Either way, they can turn our screens into places of culture shock—shared environments that can make us feel like foreigners, even when we are home.

Architectures of Loneliness

Suburbs, those physical settings closely associated with the American dream, remade the world as a set for smaller-scale shows. They were developed as responses to the incorrigible thrums of city life: In place of messy togetherness, they imposed distance. They established the home, for the middle and upper classes, as a single-family castle. People signaled their privilege by surrounding their houses with uniform outdoor carpeting: grass that had to be watered into lushness and then mowed into submission.

Suburban homes, in other words, were safe spaces before the term would become a culture-war anxiety: They rejected the city and its tumult of togetherness, imposing order on the landscape and the people who shared it. They separated people, tract by tract. They segregated people, by class and often by race. They treated both as a selling point. The fantasy on offer, in the end, was a world rid of its strangers. Each tidy homestead promised a life of synthetic ease. The people of suburbia had no need to be exposed to anything that made them uncomfortable. Life, instead, could be molded to the shape of the show. It should be seamless. It should allow you to write the lines, to call the shots, to set the scenes.

Escapism, as a way of life: The premise becomes ever more inescapable. Many of the status symbols enjoyed by the 1 percent and sold to the remaining 99 are also tools of social separation: private planes, private gyms, homes so stridently self-contained that they double as microclimates. Khloé Kardashian, Kim's sister in life and in influencerhood, recently posted a "room reveal" on Poosh, her Goop-like lifestyle website; the room being revealed was, she claimed, a pantry. In truth, it was a store, enormous and perma-stocked and resembling a retail space in pretty much every way but the checkout. The message the images sent was not merely that Khloé Kardashian is rich enough to afford a never-ending supply of Vlasic Snack'mms pickles; it was, more specifically, that her richness spared her from the indignity of obtaining Vlasic Snack'mms pickles from a public place.

Wealth as social exceptionalism is in one way a very old idea. Those who are rich have long used their money to separate themselves from those who are not; "means" has long meant the possibility of distance, physical and otherwise, from the noises and smells and movements and inconveniences of the hectic world. But as celebrity becomes more democratized, the aspiration does, as well. Everyday influencers now share their versions of Kardashian's personalized 7-Eleven, sharing massive grocery-store hauls and promoting the organization methods—shelves, bins, label-makers—they use to rein it all into sellable submission.

Why merely replicate stars' outfits or workouts or skincare routines when we could try to replicate their even greater luxury: the ability to live lives that are thoroughly set-designed? Online environments, even as they replicate cities, also replicate suburbs. The suburban promise—that main character energy could be made immersive—is an organizing principle of the internet. So is the suspicion of strangers. So is the expectation that life itself can be stage-managed into submission. Suburbia was a self-conscious rejection of urban life and its unruliness; to the extent that screens contain both settings, we navigate a place that is divided against itself. Marinate in it all for long enough, and the world begins to look not like the world at all—busy, unruly, teeming with people who deserve to be there precisely as much as you do—and more like a place that owes you its stage. *Life: The Movie*, as the journalist Neal Gabler called it, becomes *Life: The Cinematic Universe*.

Fiction creep—the encroachment of the endless stage—can lead people to lose sight of the world as it is. It can make it difficult to process facts through anything *except* our entertainment. The paradox of the spectacle is that it leads, very often, to boredom. The spectacles do not soothe us; instead, they create the demand for more: more amusement, more drama, more distraction.

Jeremy Bentham's design for the Panopticon was implemented, in Britain and elsewhere, in the penal context he'd intended. But it was also remade as entertainment. In 1850, construction began on the Royal Panopticon of Science and Art, a museum and exhibition hall and shopping center that soon became, as one magazine put it, the most "delightful lounge in London." In 1858, the building was repurposed as a theater for music and variety shows. In 1861, it hosted the debut of a trapeze act featuring Jules Léotard and his revolutionary onesie.

So the Panopticon, in Britain, turned "see and be seen" from a carceral threat to a carnivalized promise. And Americans, as is our wont, took things a step further. Visitors to Boston, today, might choose to stay at the Liberty Hotel, which transformed a panoptic prison into luxury accommodations decorated with whimsical reminders of the building's past. (In addition to a restaurant named Clink and a bar named Alibi, the Liberty promises rooms so luxurious, "we can't guarantee that you'll ever *want* to leave.")

Through similar trajectories, Big Brother, that ever-timely warning about life under totalitarianism, has morphed into *Big Brother*, CBS's ever-campy reality show. The title is apt: It acknowledges the stakes of our split-screen existence. When the comedy and tragedy share the same stage, it becomes ever harder to distinguish between them. Our version is environmental. Our Panopticon doubles as an endless theater in the round.

"Are you not entertained?" Maximus, the hero of the film *Gladiator*, screams at the Roman throngs who see his suffering as their show. The trouble is: We are.

Interesting Storms

But the broader promise of a set-designed world is that it requires very little thinking at all. Sets are places of totalizing control. They do not merely allow magical thinking; they demand it. Sooner or later, though, the spell will be broken. And we will be thoroughly unprepared. The blunt realities of the physical world can strike us as shocks—and as betrayals. The indignation can come for us individually (as when we age away from the youth we were told we could purchase for ourselves).

It can also arrive as a planetary kind of shock. Storms that grow angrier; seas that grow higher; air that gets hotter; each is a matter of physics, and thus entirely predictable. But each new hurricane or flood or quake can seem like a new kind of crisis: an interruption of the reality rather than reality itself.

A weather app recently sent a push notification offering to fill me in about "interesting storms." This is the fun fetish in action: I do not need storms to be "interesting." But when "interesting" is so regularly transformed from a description into a demand, even the weather will fall prey to the mandate.

The show must go on is fine as a principle of theater. In the wider world, though, it can lead us astray. It can limit our vision, and our agency. The planet at large has defied the old scripts; in response, we have been updating our sets. We have been meeting the crisis by turning nature into decor. Natural materials—materials, at least, designed to evoke nature as an aesthetic—have informed many of the latest trends in mass-market home design. Furnishings are constructed of rattan, wicker, jute. Houseplants, both organic and fashioned of plastic, serve as decor. Wallpapers and artwork mimic banana leaves, birds of paradise, and similar icons of botanist chic—images meant to lend even the dullest of spaces the humid lushness of the tropics.

The style is mimicked in public spaces, as well. (See: the rise of walls made entirely of plastic greenery, many of them decorated with Instagram-friendly neon light fixtures.) Climatecore, you might call it, arose during the years that found climate change shifting from a threat to a chronic crisis. It recalls the way Americans in the 1950s made sense of the space race, and the atomic bomb, by turning futurism into a design principle. It promises absolution. It hopes that, surrounded by all the manufactured evidence of a lush world, people won't stop to consider the irony that the plants on display have been constructed of plastic.

Climatecore offers false catharsis. It allows us to do nothing with such dramatic enthusiasm—such commitment to the bit—that we can feel like we're doing something. Magical thinking, on the stage, is not a delusion. It is all there is. And it is worsened when the stage holds us at a distance. "All politics is local," the old saying goes; often, though, even the most traditionally place-based elements of American life take their shape from the trickle-down effects of national culture. Arguments at local school-board meetings recite the outrages scripted every day on national talk radio and national cable news. Local newspapers, the outlets that once connected communities to themselves, are dying in droves. Many Americans would be hard-pressed to tell you who their local council members are or who their state legislators are—but able to tell you, in detail, the latest scandals involving the national government.

Pop culture channels the disconnect. Sitcoms have settings, but quite often lack meaningful locations. Placelessness is an abiding joke on *The Simpsons*: Springfield, that bit of bland suburbia, is anywhere and nowhere at the same time. *Parks and Recreation*, a sitcom that went out of its way to be specific about its setting (the fictional town of Pawnee, Indiana), was at its heart a show about national politics. Pawnee was, in the end, a microcosm of Obama-era national politics: The town had its own version of a Tea Party. Its city hall election was campaign-managed, on one side, by a savvy national operative.

Pawnee's form of placelessness is, like Springfield's, something of a joke. But it is also an insight. The national and the local, the micro lens and the macro—on the screens, they become the same thing. We still have our settings, of course: the places we live, each unique. But we come together, as a collective, in the screen. It is our shared environment. It is our collective architecture. It is a place where words and images have ever more power to shape the pictures we hold, of the world and of one another. The people we encounter, over the distance, can fall prey to similar errors of vision.

At its worst, that expectation of a sanitized world can foreclose empathy. It can trap people in their illusions, driving them to ignore the world as it is in favor of the world as they would prefer it to be. In *Regarding the Pain of Others*, that book-length meditation on the human cost of spectacle, Susan Sontag describes a conversation she had with a woman who had lived in Sarajevo when the city was in the midst of war. Witnessing an attack on a nearby area, the woman's first reaction to the horror was not shock or grief. Her impulse, instead, was to change the channel.

The Pictures in Our Heads

The journalist Walter Lippmann observed the importance of media inputs well before "media" was part of the American vernacular. His 1922 classic, *Public Opinion*, considered the information economy of the time—but managed, along the way, to anticipate the complexities of a world that cedes to its screens. Lippmann was writing not only during the early age of radio but also during a smaller kind of printing revolution: The 1920s contended with the penny press, and the new ubiquity of imagery, and the expanding influence of advertising. Lippmann was reckoning, a century ago, with the early stages of the world we are navigating today. His fear was that we could not bear the weight of all the distance. His fear, further, was that, as we tried to know one another from afar, we would become reliant on images of the world rather than on the evidence provided by the world itself. We would become addled. We would become ignorant. And we would then become vulnerable—to the images, to the advertisements, to the stories, to the overwhelm itself.

Lippmann was writing not long after Freud, and *Public Opinion* is informed by the then-nascent field of psychology. The journalist understood that the human mind is biased toward emotion over information. He understood that it tends to prefer the easy stories over the complicated ones. He understood, as well, that the mind is not always good at categorizing the information it takes in. On the contrary, all the things people encounter, by choice or by circumstance—the news stories, the novels, the Hollywood films, the Hollywood stars, the radio shows, the billboards, the histories, the satires, the amusements, the truths, the lies—would end up in the same place. They would create the "pictures in our heads," and further engagement would complicate those images, edit them, perhaps distort them.

Lippmann was also writing as American newspapers were reforming. This was the era when objectivity, as a standard, was born. It was the age when reporters instituted standards of sourcing and validation. The changes were responses to many of the problems Lippmann was pointing out: the proliferation of information and misinformation, the world-shaping capabilities of advertising, the establishment of public relations as a field and fact of political life. The newspapers were responding to market pressures by creating a new kind of commodity: information that had been collected, vetted, verified. Information that would go out of its way to clarify what had been reported and what had been merely opined. The lines were not always fully clear—nor would they ever be—but they were efforts to find new modes of order among the chaos. They were based on the idea that "the pictures in your head" are, in a direct sense, all you have.

One of the conditions of life in the modern world, the geographer Edward Relph argued, is a sense of ambient placelessness—an erosion of people's connection to specific locations that creates, in turn, an alienation from them. Relph was writing in the 1970s, well before the arrival of Walmart Supercenters and Amazon Prime and all the apps that define "convenience" as the freedom from interaction with other people. But the erosion he identified has only deepened. Local dialects are fading away as more people live—and, therefore, learn—through mass media. Local newspapers are dying away.

And localities are the basis for the pictures we hold of the world. That simple fact shapes every other. The pictures we carry around with us, in our minds' ever-growing camera rolls, are much more than representations of the world as we understand it. And the facts that feed the pictures are becoming endangered.

4
The Scripts

In the early days of radio, producers devised an array of clever tricks to add bits of audio flair to their programs. They'd use sheets of aluminum to mimic thunder, and wads of cellophane to mimic the crackle of fire. They'd mimic the sound of horses' hooves using the hollowed-out shells of coconuts.

After a while, the sound effects became so thoroughly associated with radio storytelling that audiences came to expect them. A horse, in person, will bear little aural resemblance to a pair of clanging coconuts; but a horse, on the radio, was a different matter. A radio horse demanded the clopping of coconuts. And that expectation lasted even as sound-effect technologies improved. Producers would dutifully coconut-clop even when they had the option of including the sounds of real horses in their shows. The horses had been cast into that consummately postmodern condition: hyperreality. On the radio, they were larger than life and realer than real. And once the heightened reality had become the standard, the plain version would no longer do. The coconuts remained, clop-clopping into radio history.

Today, critics talk about the "coconut effect" to describe what happens when the manufactured reality, as an expectation, overtakes the original: The producers fudge things, basically, as a matter of fan service. The concessions are typically minor. But the coconut effect can help to explain broader concessions, as well. Screens have a way of turning reality into an arms race—and when the plain truth is pitted against the manufactured one, you can probably guess the winner.

In early 2021, a mere week after a motley group of investors collaborated to inflate the stock of the video-game retailer GameStop, MGM announced that it had landed the rights to the story—not to a book about it, but to a book *proposal*. The resulting project vied for attention against the TV movie *GameStop: The Wall Street Hijack* and the documentary *GameStop: Rise of the Players*. Titled *Dumb Money*, the project was executive-produced by Cameron and Tyler Winklevoss, the nonfictional twins who were rendered semi-fictionally in the film *The Social Network*. (The then nonexistent book that would inform their production, as it so happened, was titled . . . *The Antisocial Network*.)

A version of the GameStop frenzy—a single news event, re-metabolized as not one but competing pieces of entertainment—now repeats so commonly that to be a newsmaker who isn't subjected to ambiguous fictionalization is something of an insult. Entertainment as an industry has always mined reality for its source material, but the past years have been a gold rush. The "ripped from the headlines" approach has become so common that, almost as soon as a big news event takes place, a production company will announce its plans to repurpose the event as entertainment. The headline-ripping now proceeds with Newtonian regularity. In late 2018 and early 2019, two Boeing 737-Max airplanes crashed, killing 346 people in all; by early 2020, *Variety* was announcing a "Boeing 737 Max Disaster Series in Works."

The frenzy is, as always, a matter of medium and message. The internet gave rise to streaming services, and this in turn created practically limitless space that the services' executives are eager to fill. Headline-ripping has reached its ubiquity in part because translating reality into content is, in general, more cost-effective than inventing something new. But the medium is the moral, as well—and the services market their shows, very often, as journalism by other means.

In July 2020, *The Hollywood Reporter* shared that

"Adam McKay's next project at HBO would take on the timeliest of subjects: the race to develop a vaccine for COVID-19." (The subject was so timely, indeed, that the documentary about the vaccines was announced far before the announcement of the vaccines themselves.) The following year, when Britain's SkyTV announced that Kenneth Branagh would be starring as Boris Johnson in a five-part miniseries about the pandemic in the UK, the actor brushed off criticisms that the soon-to-air project was coming "too soon": "I think these events are unusual," he said, "and part of what we must do is acknowledge them."

Without a TV show to do that acknowledging, the implication went, the unusual events—the painful events, the significant events—would go ignored. The semifictionalized treatment of the pandemic would be the first rough draft of history. It would be noble and necessary. If a tree falls in a forest and no one turns the collapse into a six-episode Netflix docudrama, does it make a sound?

The frenzied timelines can lead to awkward collisions between the shows that are *based on a true story* and the stories themselves. *The Dropout*, Hulu's treatment of the rise and fall of the blood-testing company Theranos—the series was based on a podcast, which was in turn based on a book—premiered in March 2022, as the trial for Theranos's former COO, Sunny Balwani, was taking place in Northern California. The nonfictional litigation of Theranos collided against the fictional one: Two of the potential jurors who had been selected to hear the case were dismissed; they had seen episodes of *The Dropout* and might have been prejudiced by its depiction of the events at issue in the trial.

Collisions like that grow more common as TV shows plant their flags in reality. In the early 1990s, Lyle and Eric Menendez were convicted of the murder of their parents and sentenced to life in prison without parole. In September of 2024, the producer/director Ryan Murphy retold the story of the brothers' crime and trial as part of the Monsters series he makes for Netflix. In October, the documentary *The Menendez Brothers* (also on Netflix) premiered. Later that month, the Los Angeles County district attorney's office announced its request that the brothers be resentenced—and, potentially, released.

In the fall of 2024, Murphy premiered his ambiguously fictionalized rendering of the story of Jeffrey Dahmer, the serial killer who, in the late 1970s and 1980s, murdered—and, then, cannibalized—seventeen people. In response to the series, many friends and family members of Dahmer's victims spoke out against the show that restaged the grisly killings for a worldwide audience. "It couldn't be more wrong, more ill-timed, and it's a media grab," one said. Rita Isbell, whose nineteen-year-old brother was murdered by Dahmer, described the experience of watching a version of herself depicted onscreen. "When I saw some of the show," she wrote in *The Hollywood Reporter*, "it bothered me, especially when I saw myself—when I saw my name come across the screen and this lady saying verbatim exactly what I said." (In response to her claims that her likeness had been used without her consent, *THR* reported, Netflix and Ryan Murphy Productions declined to comment.)

"Murphy and his collaborators," *Vulture* wrote in an assessment of *Monster: The Jeffrey Dahmer Story*, "are obviously aware of how exploitative it can be when the stories of serial killers are sold to a murder-obsessed public and how hurtful it is when victims are diminished, but the show never figures out a way to avoid committing the same crime. You don't get credit for lamenting the existence of a circus when you happen to be the ringmaster."

This is a defining tension among shows that engage in the tricky alchemy that turns traumas into entertainment. It remains a tension, in part, because of a broader kind of unsolved mystery: the fact that American culture, as a whole, is still navigating the difference between honoring people's stories and colonizing them. Elevation and exploitation can look similar. Often, they can *be* similar. We are still figuring out, though, where the one ends and the other begins.

The interplay between *The Menendez Brothers* and the legal fate of the titular siblings was eloquent evidence that true crime, when the screens go interactive, can double as fan fiction. In 2019, amateur sleuths helped to reveal the identities of some of the previously anonymous victims of the

Golden State serial killer. After the young vlogger Gabby Petito was murdered in 2021, self-appointed investigators, many using social-media postings as evidence, proved crucial in solving her murder. As the genres collide, though, the director's dilemma—how, when it comes to these crimes, do you balance empathy and exploitation?—applies to the new detectives, as well. When the screens are interactive, the complexities compound. When audiences recast themselves as investigators, they can bring the wisdom of crowds to the solving of crimes. They can also risk turning other people's pain into their puzzles.

In 2022, four students at the University of Idaho were murdered in the campus-adjacent house they shared. News of the killings spread; the real crime, very quickly, transformed into a "true crime"—and with all the attendant dynamics. Efforts to help bring justice to the students doubled as interactive entertainment. Facebook and Reddit groups dedicated to the task swelled to hundreds of thousands of participants. Their members shared posts that varied from the coldly clinical—analyses of autopsy reports—to the whimsical. (One post, riffing on a blind item from the gossip blog DeuxMoi, wondered whether Kim Kardashian would participate in the sleuthing.) The crowds kept offering up such assistance even after the investigators in charge of the case, claiming that the deluge was hampering their efforts rather than helping them, begged them to stop.

But the requests chafed against another tension of the

networked world: to pay attention to something is also to claim it, in a way, as one's own. The amateur detectives may have been motivated by a desire for justice; by participating in the murders' investigation, they were also benefiting from the murders themselves. They were gaining followers and fame—benefiting, whether they intended to or not, from the fickle currencies of the attention economy. "Please stop turning these poor kids into your identity," a Reddit post pleaded with the claimants. It received more than two thousand upvotes.

Category collapse can come for audiences, as well. The toppled divisions bring more questions that we have yet to answer meaningfully. Where are the lines between the spectator and the participant? What is the difference between spectatorship and exploitation? And—that quandary that plagues the producers both professional and amateur—what is the difference between telling a story and appropriating it?

"I'm not telling anyone what to watch, I know true crime media is huge rn," a man who claimed to be a relative of Rita Isbell posted to Twitter soon after the premiere of *Monster: The Jeffrey Dahmer Story*—"But if you're actually curious about the victims, my family . . . are pissed about this show." He added: "It's retraumatizing over and over again, and for what? How many movies/shows/documentaries do we need?"

The answer, for now, is *as many as audiences will watch.*

Just months after Gabby Petito's murder, Peacock premiered a documentary about her life and—much more directly—her death. The film featured an interview with one of the sleuths who had provided information that helped investigators to determine her killer. Even the woman who had helped to ensure that Petito would get justice acknowledged the costs of exposure. "Everyone wants something crazier out of this," she said. "It *has* to get crazier."

Gaslit

What the critic Roger Ebert said about the cinema—"a machine that generates empathy"—can apply just as readily to literature and TV shows and video games and the many other modes we have of inhabiting other people's lives. But it applies when the machinery itself establishes clear lines between the people on the screen and the people who actually exist. The new infotainment erases those lines and then categorizes the erasure as narrative depth. Many of its properties—deftly acted, sumptuously produced, compelling as pure entertainment—wink with such strident self-awareness that they read, in the end, as camp.

The shows tease and nod and wink—a standard thing for current TV, yes, but a striking thing for shows that, uniformly, retell true stories of exploitation, manipulation, and, in some cases, murder.

The Thing About Pam, Peacock's remake of the *Dateline* podcast that was itself a remake of a *Dateline* TV episode, investigates three murders committed by the Missouri housewife Pam Hupp (she was convicted in 2019). Narrated by the *Dateline* regular Keith Morrison, the show often offers faux-philosophical pronouncements like this: "In everyday life, we are all the heroes of our own stories. But what happens when our story gets *told*?"

Morrison's baritonic musings, in this case, are offered as voice-overs; they come as the semifictionalized Pam and her semifictionalized husband watch the *Dateline* episode about . . . yep, Pam Hupp. "Can you believe this crap?" Zellweger's Pam mutters to her husband after watching her own story on television. "Maybe no one will see it," he replies.

So here you have the campily fictionalized crime show, nodding to the campily true-crimed drama that the campily fictionalized crime show is based on. The slapstick quality suggests that audiences are not supposed to think of *The Thing About Pam* as the story of murders whose effects are endured, every day, by the very real family and friends of the very real people who were killed. Instead, we're supposed to marvel at Renée Zellweger's spot-on Missouri accent and appreciate how willing she was to don a fat suit to play the role, and maybe ponder transcendent questions about what drives people to murder. We're supposed to treat the whole thing as, basically, fiction. And that means treating the people as fictions, too.

The Crown, the Netflix series detailing the lives of the British royal family, is at once a history, a prestige soap opera, and an ensemble biopic whose "characters" are also real people. (Queen Elizabeth reportedly watched the show.) For the viewer, that collision of fact and fiction can be enticing. There's delight in wondering whether the man who is now the king of England really tried to learn Welsh that one time (he did) or whether his father really was obsessed with the Americans' moon landing (he wasn't).

Like so many entries in the genre, *The Crown* combines finicky photorealism and breezy artistic license. The series offers a stitch-by-stitch re-creation of the "revenge dress" that Princess Diana debuted after Prince Charles's infidelity came to light; it also fabricates dialogue, events, and entire characters.

The British royal family exists, essentially, to be looked at from a distance; one of *The Crown*'s resonant ironies is that its semifictional characters live out this mandate far more seamlessly than unruly humans could ever hope to. In that, the show that otherwise takes pains to display its own rank—*prestige*—is also deeply subversive. To watch it is not merely to be a spectator, but also to be a voyeur. (In 2020, the UK's Culture Secretary asked Netflix to add a disclaimer to the show that would make clear that it is, fundamentally, a work of fiction. Netflix declined.)

The tension that *The Crown*'s creator, Peter Morgan, has called a "constant push-pull" between accuracy and drama can be difficult to discern. The series appeals in large part because it presents its fictions with the swagger of settled fact. The same tension can make it confusing not just as a work of history but also as a work of ensemble biography. Are we watching people who deserve our empathy—or characters who do not?

And *The Crown* is not streaming into a world that has a terribly good grasp on the distinctions. Our world, instead, is one in which "fake news" and "post-truth" are common to the point of cliché, and where producers of all kinds offer up a bunch of semitruths and leave it to the audience to sort out the facts from the fictions, and in which the fakers and the liars will deflect blame by insisting that the fact is the thing that is lying. Fiction creep, for us, is a fact of life. *Inventing Anna*, the 2022 Netflix series about the scammer Anna Delvey, neatly sums up the approach: "This whole story is completely true," a placard reads at the outset of each episode. "Except for all of the parts that are totally made up."

One night a few years ago, my partner and I were watching an episode of *Gaslit*, the Starz series starring Julia Roberts as the Watergate celebrity Martha Mitchell. We were both side-screening with our phones, and at some point we realized we were doing the exact same thing: combing Wikipedia to find out whether the scene we'd just watched had actually happened.

In this, we were doubly chastened. First, we were spending our precious downtime . . . reading encyclopedia entries. But we were also, simply, missing the point. When you're watching a show like *Gaslit* or *The Crown*, you are supposed to accept that the story is true in a broad sense, not a specific one. You are not meant to question the difference between nonfiction and a story that's been "lightly" fictionalized. And you are definitely not supposed to be on Wikipedia, trying to cross-reference the real history against the one you're seeing on Starz.

When the French critical theorist Guy Debord described the "society of the spectacle" in 1967, he explained that the term "spectacle" would herald a moment in history when representation had replaced direct experience as people's default way of knowing the world. He might have been describing the semifictions that saturate Americans' leisure time. "What's true and what's false in X show" has become a thriving subgenre of journalism—a reliable source of clicks for the people who watch the shows and wonder. (The digital magazine *Slate* has dedicated a standalone section of its website to the category: "How Accurate Is" is going strong, at the moment, with more than 120 entries.)

But the articles rely on audiences who are willing to take the time to research the reality of the stories presented to them as pure fun. They also, in some sense, miss the point. When the past is remade as entertainment, "did that really happen?" is the right thing to wonder and the wrong thing to ask: The shows will offer no answer. They have consigned themselves to the coconut effect. Screens change history, too. They flatten it. They dramatize it. They transform the past into something that is completely true—except for the parts that are totally made up.

The muddle, you might say, is the message. And it is not new. Many of today's cultural products fit into the hybrid category that critics have long bemoaned: "infotainment." (In some ways, the compound genre was the distillation of Neil Postman's criticisms.) But their criticisms have traditionally focused on the production side of the equation: the cartoons that claim to offer education, the documentaries that double as propaganda. The complaints have applied to the consumer end of things, as well: Infotainment, being media, will have a way of conditioning audiences to expect that info, as a general rule, should be entertaining.

But "infotainment" was a coinage specifically because it was a category in need of a name. It was, in its way, novel. It was also an exception to the categories that had given structure to news consumers' expectations: "news" in one part of the paper, "arts and culture" in another. The divides were not merely matters of information and entertainment. They were also, crucially, separations between information and advertising. Newspapers and periodicals offered clear visual delineations between editorial content and advertising messages. The divides were there in screen-based mediums, as well.

The categories, of course, were never fully fixed, or fully neat. "Soap operas" were given the name because early versions of the serialized dramas were paid for by companies like

Lever Brothers, Procter & Gamble, and Colgate-Palmolive: manufacturers that saw, in the shows, new opportunities for marketing. Radio programs, the earliest settings for those dramas, were broadcast into people's homes (and, in short order, their cars and other spaces); that effectively gave the advertisements that were broadcast along with the other programming a captive audience. During the day, those audiences were primarily housewives, who listened to the melodramas as they went about their work—work that typically required the cleaning products that Lever and its fellow corporations were selling.

The soap operas provided early forms of what would later be called "synergy." Targeting their ads directly to their key consumers, the soap-sellers were solving the problem that the merchant John Wanamaker had identified in the nineteenth century: "Half the money I spend on advertising is wasted; the trouble is, I don't know which half." They were also creating the hybrid genre that would come to be known as "spon-con." The term emerged for the same rough reasons that "infotainment" did: It was attempting to acknowledge a form that, in an environment that had traditionally gone out of its way to distinguish between the editorial content and the branded, hopelessly muddled the two.

"Content," as a term, arose as a solution to all the muddle. It is meant to split the difference between news and entertainment. In practice, though, it obliterates the difference. It is a category of information that rejects the very idea that information can be categorized. It might indicate advertising or news or memes or TikTok-famous dance numbers. It might be text, audio, video. It is an all-in-one designation, which makes it useful as language but chaotic as a category. It is the informational equivalent of the "vibe": It's a designation that declines to designate much of anything at all. It's a feeling. It's a whole aesthetic. It's whatever you want it to be.

"Content" is a single word that functions like a compound one. It blurs and blends. It is a description of information that, in the end, declines to do any describing at all. But its failure as a category—its tendency, instead, to conflate and to concede—is precisely why "content" is so useful, as language. It is, in that way, akin to "story," "narrative," and other wide-net words that encompass both the false and the true. "Content" is medium and message at once. It does not merely populate the digital world; it explains it. The web unfurls in feeds and flows, streams and scrolls. Its structure, in the end, is structurelessness. And the fluidity—a dynamic that occasions what the writer Paul Ford has called "the end of endings"—is steadily sweeping away the structures that, for so long, gave information its shape.

News versus entertainment, information versus fun, commercial messages versus every other kind: These are categories that came from newspapers, books, magazines—and, more fundamentally, from the requirements of paper and print. They emerged from their own form of chaos: the one that came when people who had navigated the world primarily through words that were spoken were faced with a profusion of words that were written. They developed from anxieties both ancient and, today, all too familiar: the problems of overexposure, of information overload.

Those categories, now, are falling away. Hybridity is the way of the web. Fact and fiction are blurring, in ways that are enticing and misleading.

Facts require humility. Facts require patience. Facts require pain. Fiction requires the opposite. As entertainment grows its influence over our national stories—as it becomes, ever more, the true American ideology—it comes for the facts, too. We're seeing—and creating—a condition in which facts seem ever more fungible, in which history that doesn't fit neatly into the form of a movie seems less true than the history that does. We risk the possibility that movies and TV shows will replace textbooks as our national modes of remembering.

The past and the stories we tell about it have always been, and always will be, different propositions. When people expect even those stories to be "cinematic," though, the bigger story—the history itself—suffers.

Samuel Johnson, writing in the eighteenth century, worried that the "luxuries" of fiction would make people less tolerant of the "insipid truth." We have seen some of what can happen, already, when the 1860s are remembered through *Gone with the Wind,* or when Pearl Harbor's day of infamy is reprocessed as a treacly love story. We have also seen what can happen when the satirical version of events takes over the true one. (Many people believe that Sarah Palin said the words "I can see *Russia* from my *house*!" She did not; Tina Fey, playing the vice presidential candidate on *Saturday Night Live*, did.)

When mistakes like that happen regularly, it's a problem. When they happen regularly and nobody cares, that can easily become an emergency. The cinematic past can blur with the real one. The retconned version of the record can overtake the accurate one. Before long, history can cease to be history at all—steadily documented, commonly understood—and instead become a ceaseless act of fan fiction.

The fan service crops the uncomfortable truths from the picture. Truths, for example, like the one Hannah Arendt saw as she studied people who lived under twentieth-century autocracies. The ideal subjects of such rule, Arendt observed, are not those who have been deluded into believing lies. Instead, the ideal subjects are those for whom the distinction between fact and fiction no longer exists.

5
The Producers

In 2022, a new TikTok trend went viral. Amazon customers with Ring surveillance devices installed at their doors began asking the people delivering their packages to dance for the cameras. The drivers had little choice in the matter: When workers are treated as expendable, they are eternally vulnerable to bad reviews. And so they delivered performances along with the packages. "I said bust a move for the camera and he did it!" one person captioned her video, transforming herself into a producer and her driver into a piece of content.

Reading the reports about the trend, I did what I so often find myself doing when I take in the news these days: I stared in momentary disbelief, briefly wondered where the line was between a dystopian world and a merely weird one, and got on with my life. But I couldn't stop thinking about those drivers-turned-performers—caught in the many-Ringed circus, dancing like everyone was watching.

"The idol is the measure of the worshiper," the *Atlantic Monthly* editor James Russell Lowell once observed—and the idol, in this case, was the gilt of lucrative virality. The posters treated their posts as currency, not only for themselves but for the people who danced, their bodies slightly warped in the cameras' curves, at their command. They took for granted that being a star in someone else's movie is its own reward for a role well played. ("Lol thank you amazon driver guy! You're awesome!!!" the bust-a-move post concluded.)

This is another way that two-way screens create two-way people. When anyone can become a director, anyone might also become a bit player in someone else's show. Social media's empowering capabilities can become threats to those on the business end of the camera. Digital architectures—the swipe economies of dating apps, the exploitations of pornography—treat people as expendable.

P. T. Barnum, that great purveyor of profitable hoaxes, set the bar for turning shamelessness into an art form. (He once stitched the preserved head of a monkey onto the preserved tail of a fish—and then sold tickets for the chance to see the corpse of a "mermaid.") Barnum was a nineteenth-century showman with a twenty-first-century sense of pageantry; he understood how easily reality could be transformed from something that simply *is* into something that could be produced. He understood how much Americans would be willing to give up for the sake of a good show.

His most enduring insight, though, was not that audiences could be tricked by his fabricated spectacles. It was that they *wanted* to be. Barnum dared people to doubt him and then spun their suspicion into good PR. He paid his rivals to disparage his machinations. He bragged that "the titles of 'humbug,' and 'prince of humbugs,' were first applied to me by myself." He was a showman, and his audiences knew it: His hoaxes merely invited them to figure out for themselves how he was trying to fool them.

Audiences of today are akin to those who delighted in Barnum's spectacles. We know that "reality TV" is profoundly unreal. We are savvy to the machinery that converts footage into compelling dramas. We understand that the people who have real power in each show are the unseen producers who take the raw footage and spin it into a story.

In 2019, *BuzzFeed* published a list outlining some of the tricks that turn reality into a genre of entertainment. "17 Secrets About Reality TV Shows That'll Make You Question Everything" shared a series of modern-day humbugs, among them "lifelines for *Who Wants to Be a Millionaire* can totally google the questions" and "some couples on *Divorce Court* aren't actually married." The article did not, however, lead to much everything-questioning. It couldn't: Reality shows do the questioning themselves. They flatter their audiences in the same way Barnum flattered his: They make the trickery part of the show. They wink, yes, but also assume that their viewers will wink back.

The producers do not lie to their audiences, precisely, or try to fool them; instead, like magicians ever happy to reveal their secrets, they let viewers in on the joke. "Unscripted," most viewers are well aware, is a convenient fiction; the scenes that make their way into the final cuts are plotted and produced and edited, for the most part, into soft-lit submission. Reality stars, like their traditional counterparts, are elevated yet vulnerable. It's the producers, instead, who call the shots. Everything comes down to "the edit"; while the stars provide the footage, the edit is largely out of their hands. A 2018 *Reader's Digest* story, prefiguring *BuzzFeed*'s "question everything" post, promised to share "13 Secrets Reality TV Producers Won't Tell You"—divulged, ostensibly, by the producers themselves. The information on offer suited the subject: The producers, with a conspiratorial giddiness, revealed things that everyone already knew. One of them was this: "We're all-powerful."

Reality producers, like science fiction writers, are builders of worlds. Each show has its own physical environment (a mansion, an island, a planet reimagined as a setting for a relay race), and a defining aesthetic. Each has an argot. Each has its version of an ethical code. (A quick route to a villain edit, across several franchises, is to be accused by a fellow cast member of not being "here for the right reasons": of going on a TV show, that is to say, for the sake of being on TV.) The producers serve, in each world, as forces and first movers. They cast; they direct; they edit; they rule. "This is the true story . . . of seven strangers . . . picked to live in a house . . . and have their lives taped," announce the opening lines to MTV's long-running show *The Real World*. The most salient word, in all this, is "picked."

Reality, as both a genre and an industry, is a medium in the most literal sense: It is a conduit between realms. It creates a revolving door between the worlds of traditional entertainment and those of web-enabled influencer-hood.

Bravo's Housewives allow themselves to be edited into human tropes; in exchange, they might alchemize the exposure into musical careers or fashion lines or $100 million cocktail brands.

On *The Bachelor* and similar dating shows, the final prize is not love; it is screen time. The oversized flowers distributed in episodes' climactic Rose Ceremonies—thornless, flawless, hyperreal—offer assurances that the contestants who received them will live to see the next edit. They will be one step closer to a lucrative career as an Instagram influencer. They will retain the chance to live Daniel Boorstin's version of the American dream: to remake "personality" as a profession.

That many viewers are privy to those transactions is not

a flaw in the fantasy. It is part of the pitch. Reality, that postmodern genre in a post-truth culture, demands a willing suspension of belief. Reality shows benefit from some of the impulses that have given true crime its popularity and currency. They, too, invite viewers into forensic readings of their stories. They, too, offer mysteries to be solved. And they, too, bring their ambiguities. Are the people who star in the genre's shows people, or are they characters? Where are the lines, in a genre premised on fakery, between "viewer" and "voyeur"? At what point does consuming reality's fakeries become, on some level, a concession to them?

The questions are basic—reality TV, at its core, treats category collapse as an art form—and effectively unanswerable. Reality shows are "performative," in the most earnest ways and the most cynical. Everything, in their worlds, is performance. Their stars are performers. The scenes they create are performances, too. This is apt. Reality, in its modern form, coevolved with the internet. (*The Real World*, MTV's game-changing reality show, premiered in 1992—the year before the World Wide Web made its own public debut.) And the genre neatly mirrors the ambiguities of the two-way web. True and false, earnest and glib, highly produced and endlessly ad-libbed: Reality TV, like the internet, is a matter of both and neither. Performance, in its realm, is the answer to every question; irony, here, is not the exception but the rule. The easiest way around "Is it real?" is to put reality in scare quotes.

Mark Burnett, one of reality's megaproducers—he created *Survivor* and *The Apprentice*, among many others—tried, in the early years, to rename the genre. He campaigned to call the form "dramality," to acknowledge its mergers of drama and reality. This was one campaign he lost. But the rebrand would have done fairly little to clarify things. Whatever the genre was made *of*, the operative question was what the genre was *making*. What were these campy chimeras that were taking over American airwaves? Were they voyeurism? Were they cynicism? Were they, given their anthropological edge . . . science?

Philip Zimbardo—he of Stanford prison experiment fame and infamy—subscribed to that last view. In 2000, he praised Allen Funt, *Candid Camera*'s creator, as an entertainer who was also "a brilliant, intuitive social psychologist." After the 1973 premiere of *An American Family*, the PBS documentary series that helped to establish the reality format, the magazine then known as *The Atlantic Monthly* described the show in distinctly humanist terms. "In following the everyday dramas of the titular family," its reviewer remarked, the show was "hailed as a great breakthrough in the use of the camera in the service of knowledge."

Reality TV was all of that, but not quite. As it dissolved fact and fiction into a heady new brew, the genre repeated the pattern that so many novelties had before: Having collapsed the old categories, it found the need to establish new ones. Reality is big business. Shows once populated by phototropic amateurs are now staffed, for the most part, by people who have responded to casting calls, signing contracts and nondisclosure agreements in exchange for reality's branding opportunities. The shift has created yet another definitional problem for the industry premised on blends: Who, in legal terms, are the people who make reality? Are they workers? Performers? Artists whose medium, apparently, is artlessness?

One answer came in late 2024, when the National Labor Relations Board (NLRB) issued its decision that contestants appearing on the Netflix dating series *Love Is Blind* are employees of, rather than mere participants in, the show. The ruling entitles the cast members to, among other things, the protections of federal labor law. It hints that a long-rumored possibility could become, at some point, a plot twist: reality workers, unionizing.

The NLRB's ruling was, at heart, a matter of dull legal language: It classified *Love Is Blind* cast members as employees. But it meant that, whatever they were as figures on the screen—whether they were real to their audiences, or merely "real"—on the other side of the camera, they would simply be people, doing a job. The set would be a workplace. The ruling, in that way, marked a shift. Reality's stars, for the producers who guided them, often functioned as animated props. In early seasons of *The Bachelor*, Emily Nussbaum reports in her book *Cue the Sun! The*

Invention of Reality TV, some producers would try to manipulate cast members into on-camera breakdowns that could be repurposed as dramatic tension. Those who did so successfully were given cash bonuses and other rewards.

Some former show participants, for their part, have broken their NDAs to accuse their shows' producers of exploitations that edge into abuse: sleep deprivation, food deprivation, the overserving of alcohol, and more. The experience on the set of *America's Next Top Model*, a former contestant alleged to *Business Insider* in 2022, was "psychological warfare," and many others have come forward to make similar claims about the show's treatment of its participants behind the scenes. But evidence of mistreatment, on *ANTM* as in many other shows, also makes it to the final cut. Reality cast members are regularly shown stumble-drunk or vomiting or caught in emotional breakdowns. Were such scenes presented in straight-ahead documentary form, they might spark viewers' sympathy. They might encourage concern. They might make one wonder about the difference between spectatorship and complicity. They might, at the very least, give pause: *Is this really . . . entertaining?*

But the scenes are not packaged as reality. They are packaged as "reality." And the air quotes, allegedly, exempt the shows' viewers and producers alike from the need to wonder about the line between watching people and exploiting them. The show, for reality stars, is governed by the values—"values"—of entertainment itself. Whatever might happen behind the scenes lives in the same amoral abyss. So does, indeed, whatever might be aired in plain sight. Questions about mistreatment can apply only to people who are real. And the figures who perform "reality," the shows have typically insisted, are not real—not in any way that audiences need to be concerned with. The screen offers them a chance at fame, and fame rationalizes everything else, for producers and viewers alike. When people are there to entertain you, it doesn't matter whether they're enjoying the show. All that matters is that *you* are.

Very Special Episodes

There is a poetic kind of irony in the fact that *Love Is Blind* was a catalyst for potential improvements in reality casts' working conditions. The series began streaming on Netflix in 2020—its popularity was likely boosted by its coincidence with the early days of the Covid pandemic—and has since expanded into multiple seasons and spin-offs. It is a typical competition-based reality show that does everything it can to insist that it is not your typical, competition-based reality show. Instead, *Love Is Blind* insists, it is "a social experiment." The series is also, even more than its fellows in reality-assisted heterosexual romance, a piece of entertainment that doubles as a morality play.

The series casts a group of thirty or so singles—half of the group women, half men—who have tired of "traditional" dating. ("Traditional," on this show, includes dating apps.) The show's core feature is architectural: a small room, windowless but outfitted with a couch and some offhand decor, that the show calls a "pod." Each space features an audio connection to another pod—and this is where the show's dates (and, sometimes, the romances/engagements that spin from them) take place.

The experiment, here, revolves around the question of whether people can fall in love with each other without seeing each other. Or, as the show's cohost, Vanessa Lachey, asks multiple times each season, each time with equal urgency: "Is love . . . truly blind?" As it attempts to answer the question, the show becomes two things at once. On one side, it's "science," of a sort: an orchestrated experiment that is meticulously constructed, cast, and controlled. Cast members live, at first, in shared (but gender-segregated) living quarters, with no TV, no "devices," no outside influences—a typical feature of reality competitions, but one that, here, serves the premise of "experimental" purity.

But the experiment expands, as each season goes on, from the pods and into the "real world." The couples who get engaged in the pods (engagement, here, is required if you want to move forward in the "experiment") embark on a series of steps meant to mimic a traditional relationship: They meet in person. They go on a vacation together. They move in together. They meet each other's families. The timeline is condensed—the whole thing plays out over a month or so—but this is treated as a feature rather than a bug. Reality TV has long channeled Americans' dreams of getting rich (and thin, and cured, and beautiful, and loved) quick. *Love Is Blind* promises that "get married quick" was equally attainable.

Love Is Blind deploys the wan maxims of the typical reality show—its stars regularly wonder whether their fellow contestants are "here for the right reasons"—but it expands its purview to moralism of a more conventional strain. Cast members speak, often, about self-love, self-respect, partnership, compromise, forgiveness. They are loving like everyone's watching. They discuss, with earnest detail, their convictions about what romantic love, overall, "should" look like. They make copious use of the argots of therapy. They talk about generational trauma. They talk about personal trauma. They talk about the need to "be vulnerable." They talk about their desire to "live their values." They make giddy proclamations: "I've never felt like this." "Y'all, I just met my *wife*!" "He's my *person*."

Almost never, in any of this, is the obvious acknowledged. *The Bachelor* and its spiritual siblings, as much as they claim to put the "true" in "true love," are also blunt about the fact that they are TV shows. The whole thing, from initial meetings to Rose Ceremonies to Fantasy Suites, is unapologetically overproduced. But *Love Is Blind* asks audiences to suspend disbelief. The show operates with a documentary-style sobriety, its unflinching eye exposing cast members as they fight and make out and put on sleep-apnea masks and have panic attacks. It's not voyeurism, each scene assures, if you're watching a science experiment.

But the illusions are shattered in each season's finale, when the high-speed courtships lead to real-life weddings: bridesmaids, groomsmen, parents, friends. The weddings do not always result in marriage; instead, the altar is the place where the findings of "the experiment" are revealed. (The officiants, across the ceremonies, utter the same line: "Now is the time to decide if *love* is *blind*.") Either both people say "I do" . . . or someone is jilted before their gathered loved ones and a worldwide audience. If the latter, the show's cameras will typically pan across the failed union's gathered guests, capturing the tears of mothers who have not been made mothers-in-law, or the rage of protective brothers who have just watched their little sister say "yes," only to receive a "no" in reply.

In the lead-up to the climactic answers, the show's editors make use of nearly every trick in the reality-producer's handbook to pack suspense into the proceedings: cliffhangers, melodramatic pauses, heartbeat-style sound effects.

Those shots of the cast members' families, though, obviate the need for manufactured spectacle. The real people who have been snared into the show provide more than enough drama. They are the ones who give the lie to the show's pretensions to moralism. They are the "involuntary amateurs" that John Crosby was referring to when he warned about *The Candid Microphone*. They are the collateral damage of the love that comes from "reality."

#PlaneBae

The setting: an Alaska Airlines flight from New York City to Dallas. The key players: four people who had been seated near one another on the plane. The drama began when a couple who wanted to sit together, Rosey Blair and her boyfriend, Houston Hardaway, asked the woman next to Hardaway whether she'd be willing to do a seat-swap. The woman (her name, the couple would later learn, was Helen) agreed. As she moved to the row just ahead of the couple, they thanked her with a joke: Maybe her new seatmate would end up being the love of her life.

This would typically be the end of a story. But it turned out that Helen's new seatmate (his name, the couple would also soon learn, was Euan) was good-looking. Which was convenient, because Helen herself was good-looking. The strangers began talking. And they kept talking for the duration of the flight—about their jobs (they were both fitness instructors, it would turn out), their views on marriage, their feelings about kids, the reasons they were both still single. He showed her a picture of his mom ("So hot!" Helen gushed). Their arms, from their shared seats' armrests, occasionally brushed each other. At one point, she seemed to be resting her head on his shoulder.

The two seemed to be living out the gauzy fantasies of the rom-com. "If you look for it, I've got a sneaky feeling you'll find that love actually is all around," Hugh Grant's lovelorn prime minister says in *Love Actually*, as he describes people greeting each other in airports. Helen and Euan seemed to be finding it while aboard the airplane itself. Their meet-cute was a "Modern Fairy Tale," NBC's *Today* show would later report.

But the reason we know about Helen and Euan in the first place is that Rosey and Houston, their inadvertent matchmakers, turned their story into content. Overhearing the strangers' conversation, they soon began documenting it through pictures and videos, captured from the space between the seats. The next day, Blair shared the whole saga on Twitter. "Last night on a flight home," she wrote, "my boyfriend and I asked a woman to switch seats with me so we could sit together. We made a joke that maybe her new seat partner would be the love of her life and well, now I present you with this thread."

Blair's story, narrated as if it were happening in real time, became a sensation—retweeted by more than three hundred thousand people, liked by nearly a million. People posted videos of themselves reacting to various twists in the narrative. They commented on the tale at hand and speculated on the ones that might follow. They wondered about the couple's on-the-ground identities. They identified them, in the meantime, as hashtags: #PlaneBae and #PlaneHunk.

The online mob, in this case, wanted a happy ending for the people in its purview. And the story seemed to provide it. Helen and Euan seemed to leave the airport together—a conclusion evidenced by the photo Blair posted of the pair's retreating forms as they walked toward baggage claim, their hard-shell carry-ons in tow.

Aaaaaand . . . scene, you might say, of the potential new couple's heartwarming story. But the scene didn't end because the story didn't. Instead, as #PlaneBae spread, more and more people began talking about it—and, in the process, laying claim to it. #PlaneBae truthers attempted to puncture it. #PlaneBae believers found the identities of Helen and Euan: their employers, their family members, their social-media handles. (In response, Euan added "Plane Bae" to his Twitter bio, along with a new hashtag: #CatchFlightsANDFeelings.)

National media outlets reported the story as news. Comedians used it as fodder. The meet-cute became a meme and then it became media, with "eyeballs" glued onto it and ads sold against it. It became content and then a commodity. The lines between telling a story and selling it faded, ever more, in the frenzy. Rosey Blair, revealed to be a comedian and actor, cheerfully acknowledged the benefits of virality: She applied, via tweet, for a job at *BuzzFeed*, offering to write content for the site. (Euan Holden retweeted the post.) She invited her new fans to write a screenplay with her. She appeared on morning shows eager to share the "modern fairy tale" with their viewers. Holden appeared with her, effusing about Helen. "She's a very, very, very lovely girl," he said on the *Today* show—and one, he added, who "has a lot to say for herself."

The audience, however, had to take his word for it. #PlaneBae was a story told about Helen, rather than with her; the story's growth, too, came without her participation. The *Today* show, anticipating viewers' questions, explained in a brief voice-over why the tale it had been telling lacked its appointed damsel: "Helen preferred not to be interviewed for this story."

The #PlaneBae saga was a new twist on an old story. It is an outgrowth, in some ways, of the voyeuristic culture that rose in the 1990s, when a series of technological changes—among them the rise of cable news, with its thirst for twenty-four-hour programming—turned the already-flimsy notion of stars' privacy into an oxymoron.

"We tell ourselves stories in order to live," Joan Didion wrote so famously. The web, though, can shift the proposition. It is eliding the difference between telling stories and selling them. A few years ago, after my partner booked a hotel room for a weekend trip, the confirmation email noted that the stay would allow him to "craft your next story"—a stay at the Courtyard by Marriott, apparently, transforming him into his own life's auteur. An email I received from TurboTax informed me, cheerily, that "we've pulled together this year's best tax moments and created your own personalized tax story." Nothing is certain now, apparently, but death, taxes, and the fact that my Form 1040 will double as a biopic.

Americans are now being marketed the ability to market our stories. If the broadcast technologies of the twentieth century created a society of "performing selves," as Warren Susman argued, the two-way technologies of the twenty-first are adding an element to the equation. We are also, now, storytelling selves, mining life for stories that might become content.

The Manufacture of Dissent

In a 2015 study, Walter Quattrociocchi, a professor of data science at Sapienza University of Rome, and several colleagues analyzed more than fifty-four million comments made in Facebook groups over a four-year span. They found a disturbing and abiding trend laid out in a paper titled "Emotional Dynamics in the Age of Misinformation": The longer a discussion goes on, the more extreme people's comments become. The more polarized; the more extreme; the more cruel. The rule holds across topic areas. It takes on, in that way, a Newtonian inevitability: In the web's laws of thermodynamics, heat renews itself so reliably that, given enough time, it will make a fire.

The physics at play in all that are not natural. They are, instead, thoroughly man-made. The trajectories Quattrociocchi et al. described—inertias made incendiary—are the products of decisions made by engineers and executives but informed by corporate boards, investors, and the belief that a company's highest aspiration should be the growth of its bottom line.

And those decisions, in turn, defer to a metric that, once encoded into the algorithm—and once enshrined as a goal—becomes its own form of physics. "Engagement," in the environs of social media, is not merely attention. It is attention that proves its worth through interaction—whether likes or loves or shares or shouts. And it is attention that transforms the performance mandate that Warren Susman identified into an algorithmic mandate.

"Social media is a mini narcissism engine," the scholar Peter Pomerantsev writes in his 2019 book *This Is Not Propaganda: Adventures in the War Against Reality*. Algorithms that flatten attention in this way (he cites YouTube's as an example) also end up "encouraging us to take ever more radical positions." In digital discourse, Pomerantsev notes, "you're not trying to win an argument that plays out rationally in a neutral space; you're simply trying to get attention from more and more people." The cycle is elegant: The conversations drive polarization, which in turn drives up the demand for content that serves polarization.

The outrage on the screens, soon enough, becomes part of us. Constant exposure to stress—anger, shock, fear—can result in the shutdown that is sometimes called "amygdala hijack." The brain is endlessly adaptable: If it spends enough time being angered or terrified—caught in the endless acuity of fight-or-flight—the heightened emotions, after a while, can become baseline ones. The mind that experiences endless drama will come to expect endless drama. And then it will come to crave it.

The phenomenon helps to explain why, on TV, "news" often means "a group of people yelling at one another." And why, online, the attention economy is bolstered—and shaped—by an outrage economy. Through roughly the same mechanics that make "FLAVOR-BLASTED" foods so endlessly enticing, the information we consume comes to seem more palatable when it is artificially flavored. And when the junk becomes the diet, the health effects are similar, too. An excited brain, when the excitement never ends, becomes an addled one. It expects too much of the world. And the expectation, over time, becomes a failure of vision. The mind, once hijacked, confuses the important things for the merely dramatic ones.

"Stay tuned," the old TV shows used to say. "Don't

touch that dial." The shows' writers wove those requests into their storylines: They saved their dramatic reveals for the sections of their series that aired after the commercial breaks. Cliffhangers were created not just as means of melodrama, but also to ensure that the audiences who had tuned in one week would stay tuned in the week that followed. The world itself resists such ploys: It is the ultimate unscripted drama. It has its own schedules, its own arcs, its own plot twists. But Americans have become conditioned to expect that the chaos of human events will be edited into watchable submission. We have come to assume that the world will do all it can to save its audience from that most commercially unfriendly of conditions: boredom.

Extremism is typically discussed in political terms, as discourse brought to its (il)logical conclusion. But extremism can come in many forms. It can also become banal. The typical complaint against cable news is that it rewards performative emotion at the expense of reasoning, logic, and respectful debate. But cable is nothing compared to what discourse looks like on the internet.

Screens, just as they flatten people, can flatten our interactions. People don't have mere opinions; they have "takes." And conversations in the take economy often resemble negotiation: arguments made not in good faith, but as starting bids in a broader transaction. You might call this the "debate me, bro" mode. And it is incentivized by the algorithms that prioritize extremism—that write the double valence of "outrageous," as anger and as spectacle, into their equations.

The problem is that the algorithm operates backstage: an ethereal Oz, pulling strings of 1s and 0s. Its operations aren't legible to people participating in the conversations. To them, the angry exchanges simply look natural: just how people engage with each other. The highly produced reality show can seem like a documentary. The whole thing both heightens extremism and normalizes it. The Overton window opens ever more widely—and then yells at you for not being a door.

The political scientist Daniel C. Hallin proposed a model of discourse that came to be known as "Hallin's spheres." There is the sphere of acceptable discourse: the areas of obviousness, where ideas are held by consensus—and, thus, not in need of debate. There is the sphere of arguable discourse: the area held for ideas that are debatable but within the realm of acceptability. And then there are the spheres of deviance: the ideas so far beyond the pale that they do not deserve to be dignified with debate.

The outrage economy moves the spheres ever closer. The Venn diagrams overlap ever more closely—until, finally, they are a single circle.

Extremism, along the way, becomes an equal-opportunity proposition: a feature of everyday discourse so common that it is nearly a cliche. To like something is to be "obsessed." "I'd die for her," people might say. (Or, sometimes: "I'd kill for her.") The jargon is not literal (one hopes), but is instead a matter of digital logistics: The algorithms that drive social-media platforms tend to reward both raw emotion and rhetorical excess. And the physics of the screen mean that we are experiencing other people at both extremely close range and extreme distances.

Any screen can also have a distance-enforcement mechanism. It takes the fleshy complexities of life and flattens them into two dimensions. Even on the social-media platforms explicitly premised on conversation—X and its counterparts—people very often talk *to* each other and *about* each other, while declining to talk *with* each other. Subtweets and screenshots—features that allow people to talk behind each other's backs, before a potential audience of millions—are particular ways to use those platforms; they also give new ease to (non)engagement. Early analysts of the web, in a fit of ecstatic optimism, declared that "the internet runs on love." They were wrong, it would turn out. On many services, it runs on passive aggression. Digital forums, panopticons that arm their inmates with retweet buttons, bring extremes of vulnerability, too. They are reminders that the attention economy has no opt-out button.

Americans typically think of attention as a relatively straightforward proposition: a thing people have and a thing people give. We pay attention; it is our most obvious and intimate currency.

But the reality is more complicated than the language lets on. The attention economy may imply fair trades within a teeming marketplace, people empowered as life's producers as well as its consumers. But the attention economy does not operate in a free market. Instead, it tends to impose itself on us—in ways that make us profoundly vulnerable.

"My experience is what I agree to attend to," the pioneering psychologist William James observed. James considered attention to be an agreement, of sorts, between one's mind and one's environment. We can give attention—*pay* it—and revoke it, at will. His framings of attention, made in the late nineteenth century, still inform the way Americans talk about attention today. They do so despite the obvious: William James did not spend time on the internet. On its screens, so much competes for people's attention—so much imposes itself, breaking-newsing and push-alerting and beeping and flashing and popping and buzzing—that to pay attention is also, very often, to long for a refund.

But James, spared from the anxieties of "screen time," intuited screens' workings, as well. Attention, for him, doubles as a field of vision. The mind, like a camera lens that can never be capped, is ruthless in its focus: It has one setting and one speed. It crops every picture, automatically. Because of that, it turns attention into a constant choice. To look at one thing is also to *not* look at everything else.

Attention, in James's framework, is a zero-sum proposition. Experience, as the product of our attention, is zero-sum, as well. It is the product of the minute-by-minute choices we make for ourselves: decisions about what to see, what to read, what to learn, whom to listen to. Not every decision will be wholly our own—we might attend to our algebra homework or tax returns even if we have no bone-deep yearning to do so—but every decision will be just that: a decision. Attention is an act of endless curation. Experience, the hard evidence of our attentional choices, is shaped as well by all we choose to ignore.

Attention, in a direct sense, is everything we have. It is our most priceless possession; it is our most valuable gift. For the architects of our digital spaces, however, it is simply a commodity. Attention is "the world's most endangered resource," the journalist Chris Hayes argues in his 2025 book *The Siren's Call*; on screens, it is extracted from us. It is more specifically, as the Columbia professor Tim Wu puts it in 2016's *The Attention Merchants*, "harvested."

The metaphors are aptly visceral; we users, as the fields and the soil, are rarely aware of the reaping. We might get hints of it, maybe, while rapt in an endless doomscroll, our brains both buzzing and numb. We might be aware that, as the saying goes, *if you're not paying for the product, you are the product*. But the transformation—a person, becoming a product—is mostly imperceptible. We look at what we are meant to look at. Our movements are measured, tracked, analyzed. We live, Hayes writes, within "a kind of attentional warlordism." Our gaze is a battle ground for conflicts we'll never win. Our attention, ever on auction, goes to the highest bidder.

There's No Such Thing as Bad Engagement

Engagement as a metric is something of a misnomer. The term might imply community, togetherness, the beneficial elements of discourse. Adopted for algorithmic purposes, though, "engagement" makes little distinction between love and hate, between conversations and screaming matches. When engagement is the metric, questions about benefit and harm, social good and social ills, cannot apply. Engagement, in that way, is tautological.

You might also say that it is amoral. The calculations that lead to all this are fairly straightforward—"engaged" audiences are captive audiences are lucrative audiences—but the assumptions informing them are slightly more complicated. "Engagement," before it became a matter of infrastructure, was a matter of marketing. It described the way people would interact with commercial messages and carried a whiff of the old logic: There's no such thing as bad publicity. The term was value-neutral, and in that very much akin to similar metrics used to assess audience sizes and ad rates: "impressions," "eyeballs," and the like.

In *This Is Not Propaganda*, Pomerantsev argues that, with these assumptions, the platforms are not merely defining attention. They are also defining desire. They are assuming that the two are the same thing when, very often—especially in digital environments—they are entirely different propositions.

Americans often discuss the political future in terms of "guardrails." Taking for granted the old truism that those who achieve power will be incentivized to keep getting more of it, they emphasize the need for balances—and, even more important, for checks. Holograms are tidy—and surreal—reminders that technology and culture, when left to their own devices, will often adopt a "just because they can" approach to the worlds they are building on our behalf. "Move fast and break things" was for a long time Facebook's de facto operating principle. "Ask for forgiveness, not permission" remains a widely held tenet in Silicon Valley and its counterparts.

When John Suler outlines the disinhibitions of online engagement, what he is really describing is a category error that can rationalize cruelty: The people on the other end of the screen simply don't seem real. Suler is explaining, with elegant efficiency, one root of our many crises: the bullying. The loneliness. The insistence that character is best understood as "character."

He is also suggesting the expansive power that technological platforms have—not just to build digital spaces but also to shape our human infrastructures. Architecture guides people into action or inaction, influencing the ways people encounter one another or fail to. Social-media services do similar work, but at an unprecedented scale. They are mediums with implicitly moral messages. In their ecosystems of 1s and 0s, they contain us. And, for better or worse, constrain us. They skew our vision of one another. They take the still-unanswered tensions of reality TV and its corollaries—the terse transactions of seeing and being seen—and make them environmental.

The tech platforms, that is to say, operate very much like reality producers. They are directing things from backstage, we know. We are, we further understand, players—actors, extras, scriptwriters, human props—in their shows. "Reality" being what it is, though, the rest is often illegible to us.

We tend to be, however, extremely legible to them. This is, of course, by design. When the goal is lucrative melodrama, imbalances are features, not bugs. A producer who knows performers' triggers and tics can serve up ever-more-bespoke manipulations—and in that way serve the show. In 2015, Stephen Hawking interviewed Mark Zuckerberg before a crowd at Facebook's headquarters. The physicist, playing to type, asked the executive about universal theories: What big problems and questions did Facebook's founder want to solve?

Having given his answer—*the mind and its mysteries*—

Zuckerberg offered one more. "I'm also curious about whether there is a fundamental mathematical law underlying human social relationships that governs the balance of who and what we all care about," he said. "I bet there is."

Whether humans can be reduced to numbers in the literal sense, we live in worlds that are engaged in everyday acts of reduction. Our tastes, our friends, our likes, our resentments, our purchases, our aspirations, our interactions—the stuff that, in another era, suggested the giddy complexity of being human—now have their digital doubles. They are data, stored in drab-walled server farms studded across the landscape. They turn us, in some sense—in the all-seeing eyes of the platforms—into data.

They do that, of course, in the name of customer service. They do it in the name of predicting—and, then, responding to—our incorrigible desires. "Love was invented by guys like me to sell pantyhose," the fictional ad man Don Draper remarks in *Mad Men*. The new marketers turn the invention back into data. They treat us as advertisements for ourselves. They achieve newer versions of what the pioneering ad man, Edward Bernays, observed about the marketers who would shape so much of Americans' conception of themselves:

> Those who manipulate this unseen mechanism of society constitute an invisible government which is the true ruling power of our country. . . . We are governed,

our minds are molded, our tastes formed, our ideas suggested, largely by men we have never heard of.

... In almost every act of our daily lives, whether in the sphere of politics or business, in our social conduct or our ethical thinking, we are dominated by the relatively small number of persons ... who understand the mental processes and social patterns of the masses. It is they who pull the wires which control the public mind.

Bernays made those observations in a book that was titled, aptly enough, *Propaganda*. But he took a relatively sanguine view of the possibilities of public persuasion, and with roughly the same rationale that our digital reality-producers use: The total control he describes was enforced in the name of commerce. We cede our minds to marketers, in Bernays's framework, so that we might benefit from all the goods they have to sell us. We give up a little of our individual freedom so that we might avail ourselves of that quintessential American liberty: the power to buy our way into better lives.

This is the transaction that many tech platforms enforce, as well. We give them—*pay* them—our attention; in exchange, they promise that they can come to know us better than we know ourselves. The trade-off has led to some occasional awkwardness (as when digital retailers, parsing their customers' micromovements, have served ads for pregnancy products before the pregnant people themselves learned the good news). For the most part, though, it's an exchange Americans have accepted. We take for granted that the observation made several years ago by the University of Virginia professor Siva Vaidhyanathan will retain its currency: "So far, Google manages us much better than we manage Google."

Whether it's Google or Facebook or Instagram or TikTok or X or one of the many other platforms that will follow, they will very likely follow the logic that has governed so many technologies before: "We shape our tools, and thereafter our tools shape us." They will also, likely, follow the course that so many technologies have taken before, transforming in short order from opt-in services to default expectations.

Americans often talk about "the algorithms" in the same rough ways that ancient people used to describe their gods: hovering, mysterious, vindictive, all-powerful. We accept, just as they did, that humans must sacrifice something of ourselves if we hope to remain in the powers' good graces. Daniel Boorstin's predictions about the arc of American life—that it would involve images so seemingly real that we could simply inhabit them—have taken their form, now, in the formless expanses of the web. If residence in those spaces means that "our minds are molded" by anonymous architects . . . that is simply the cost of doing business.

As the machines reflect us back to ourselves, though,

they also shape us. They instruct us. They goad us. They absolve us. In a scene in the comedy series *Fleabag*, the title character—a decent character who often acts like a bad one—strikes up an unlikely relationship with a young Catholic priest. Their relationship is ambiguously defined: They go on outings that seem like dates; they banter; they flirt; she begins referring to him as "Hot Priest." The uncertainty of it all culminates in a late-night, drunken rite: He hears her confession.

She shares a lifetime's worth of misbehavior, waiting to hear gasps of shock from the other side of the confessional booth. She hears none. And then she rids herself of the thing that has truly been weighing on her. "I want someone to tell me what to wear in the morning," she says. Preempting Hot Priest's objection—what a normal thing to want—she clarifies:

> I want someone to tell me what to wear *every* morning. I want someone to tell me what to eat. What to like, what to hate, what to rage about, what to listen to, what band to like, what to buy tickets for, what to joke about, what not to joke about. I want someone to tell me what to believe in, who to vote for, who to love, and how to tell them.

Fleabag (the show never reveals her given name) is the type of thoroughly modern woman who is used to wrapping herself in protective layers of irony, humor, distance. The lines find her being—with another person, with herself, with her audience—more honest than she's ever been in the show. It is not, strictly, a confession, because it is not, strictly, an admission of wrongdoing. Instead, it is a speech about feminism. She has benefited from its fights—she lives its freedoms—and in some ways resents them. They foreclose passivity. They demand that she be her life's own writer, producer, actor, audience. They turn living, itself, into endless work. They turn identity from something one simply is into something one must constantly choose.

Her brief monologue is tinged with the sometimes wearying freedoms of life in the modern world. It is also tinged with melancholy. Here she was, after all, faced with the upshot of her aching admission: She is craving something that doesn't exist.

But the melancholy is laced with irony. This character lives in the same world we do—one where so much, from our smallest decisions to our biggest dreams, are being decided on our behalf. Her monologue finds her, in that way, trapped in that quintessentially modern form of despair: desperately craving something she already has.

Screens can replicate, in that sense, the medieval European festivals known as "Abbeys of Misrule." The carnivals were events that briefly inverted the social structure, allowing peasants to act as lords and ladies, mocking them in the process. The performances looked like acts of subversion. But what they amounted to, historians suggest, was complacency in the guise of fun. Like pressure cookers releasing their steam, the days of sanctioned satire allowed people to vent the frustrations of life within a rigid hierarchy while doing nothing to change the status quo. The festivals' claim was catharsis; their broader effect, though, was stasis. Their outrages cleansed, the peasants went on with their lives, their low status intact.

Screens host daily versions of that ritual. Bringing as they do the possibility that our images and words might be broadcast to millions of people, can give the illusion of political agency. What they offer instead, all too often, is a faint facsimile of power. They turn political agency into an aesthetic: something we might appreciate, from a polite distance; something we might feel rather than do. Audiences are empowered only to watch and to react. They can clap; they can heckle; they can leave. The structures of social media—the narrow economies of reaction they offer to their users—mimic in large part those limitations: clap, like, log off. Under the guise of empowerment, the platforms often assume passivity and encourage it. They entice. They lull. They can turn citizens into passive audiences.

On May 24, 2022, nineteen children and two of their teachers were murdered at Robb Elementary School in Uvalde, Texas. One of the many revelations made in the aftermath of the tragedy was that the massacre might have been prevented. In online messages sent before the attack, the man who would commit the massacre had threatened to rape, kill, and kidnap teen girls on the streaming and messaging app Yubo. The girls reported the threats. They received no reply. As one of them explained: "That's just how online is."

6
The Extras

In 1981, the Harvard Law School professor Roger Fisher published a brief article in the *Bulletin of the Atomic Scientists* under a heading both hopeful and bleak: "Preventing Nuclear War." The idea, part proposal and part thought experiment, went like this: The "nuclear football"—the briefcase carried by a military aide, always in the president's vicinity—would no longer contain the codes and other pieces of documentation required to order a nuclear strike. Instead, it would carry a butcher knife. The codes themselves would be surgically implanted in the chest cavity of a volunteer, in a location very close to the heart. To access the codes, the president would need to retrieve them, gruesomely and fatally, from the body of another person.

The moral logic of Fisher's idea was blunt. If you are willing to kill many people from a distance, you must also be willing to kill the one person who is standing right in front of you. He was adapting classic conundrums of moral philosophy—the trolley problem and its many brethren—to the grim capabilities of the nuclear age. The system, he argued, would force the president to "look at someone and realize what death is—what an innocent death is."

Fisher's proposal was never implemented. But it remains potent as a thought experiment. Physical distance can amount to moral distance. That is true in theaters of war and in the theaters of everyday life. It is, in general, easier to speak badly of people behind their backs than to their faces. It is easier to be cruel to people who manifest as images and avatars rather than right-there-next-to-you people—flat figures who exist within the pixels of the screen.

The contemporary social psychologist John Suler describes what he calls the online disinhibition effect: the tendency for people in digital spaces to act in ways they never would in those spaces' offline counterparts. Suler attributes the effect to several misapprehensions those people might carry, among them solipsistic introjection (the sense that the online world is "all in my head") and dissociative imagination (the sense that the online world, even if it's real, is merely a stakes-less game). The core misapprehension, though, is the assumption that the digital environment is fundamentally different from the physical one. The error extends to the people one encounters in that environment: They may look like people. They are "people."

The disparity has a moral valence, too. One person, right in front of you, can seem to make moral claims that even thousands of people, seen from afar, cannot. "A single death is a tragedy, a million deaths are a statistic," Stalin said. The observation explains his own failures of his vision, but it also hints at our own. Fisher's thought experiment has found a grim new currency: a relic of the nuclear age, made urgent in the digital.

"Your right to swing your fist ends where my nose begins," the old line goes. But when the other person's nose appears as a picture on a screen—when it looks, in a quite literal sense, imaginary—the principle loses its simplicity. The old moralities no longer apply. Are we interacting with people or pictures? The answer is "yes." And that can leave us flailing when it comes to the most elemental of questions: How should we treat one another?

Online disinhibition radiates. And it is abetted when *all the world's a show* becomes a premise of digital life. Characters are, above all, expendable. Their purpose is to serve the story; when their service is no longer required, the brute cosmology of the TV show assumes, they are ready to be written off. The assumption now extends to the real people who play their parts in our uncanny theater: Once people stop being delightful, some assume, they should simply stop being. They should be pulled, forcibly, from the show. They should be "canceled." It has now become common to call for leaders to be impeached as soon as—and, sometimes, even before—they take office. But impeachment, in the land of the screen, is a bipartisan possibility. To those we disagree with, we might say, "I disagree." We might, though, cut to the chase: "Just log off." "Delete your account."

Distance can alienate. Distance can rationalize. On the internet, it can be overcome in an instant. At alarming scales, and with alarming ease, we can now wound people we've never met. We can do it intentionally. We can do it unthinkingly. On our screens, we bear both the power of Fisher's president and the vulnerability of his potential victims. But we also bear the uncertainty that comes when we're never quite sure who and what to believe. Suler's framework is stark: If the digital environment isn't real to you, the people who populate it won't be, either. The people we meet on our screens are right there in front of us, like Fisher's presidential aide. They are also, like Fisher's distant victims, not there at all. Proximity, in that way, becomes a distinction without a difference. They are all real. Anyone can be harmed when the bombs start to fall.

Poetic Faith

On December 4, 2024, Brian Thompson, the CEO of the insurance giant UnitedHealthcare, was fatally shot on the streets of New York City. On December 9, police apprehended a twenty-six-year-old man at a McDonald's in Pennsylvania, charging him with the murder. His name, it would soon be revealed, was Luigi Mangione. And he became, despite and perhaps because of the crime he was accused of committing, a folk hero. After his arrest, people online began exploring the postings Mangione had left in the life he led before—on Twitter, on Facebook, and, in particular, on the book-sharing site Goodreads. He was, it seems, a prolific reader. The titles Mangione had listed on the site (*The Lorax*, *Atomic Habits*, *The Four-Hour Workweek*, and *1984*, among many others) offered a range that seemed to defy easy interpretation—and, some suggested, sense.

"His posts had something for everyone," the news site *Axios* observed: "Conservatives saw an anti-capitalist San Francisco liberal; progressives saw an 'anti-woke' rich kid who aligned with right-wing futurists." The culture critic Mark Harris, posting on Bluesky, called him an "ideology tourist." The writer Max Read suggested that the seeming incoherence was actually, in its way, familiar: "It's a loudly non-partisan, self-consciously 'rational' mish-mash of declinist conservativism, bro-science and bro-history, simultaneous techno-optimism and techno-pessimism, and self-improvement stoicism—not left-wing, but not (yet) reactionary, either."

The reading list itself became its own text to be read

and analyzed. And it was, particularly against the tragic backdrop, notably poignant. The impression given, by these books both read and aspired to, was of a person who thirsted—not just for knowledge, but for meaning. Mangione (or, at least, the person he was when he created the posts) wanted to understand the world so badly, it seems, that he made a point of continuing his education well after he'd earned his degrees. He cobbled together a syllabus, of sorts, and an array of professors by proxy: Dr. Seuss, George Orwell, Timothy Ferriss. He shared his lesson plan with the public. He seemed to be using books, those ancient agents of wisdom, to feel his way through messy modernity.

But: *seemed to be.* These were social-media posts. The analysis of his reading habits, as fervent and detailed as it was, was also speculation. But that was part of the point. Mangione was in jail as his book preferences were being dissected. He could not speak for himself. He could not correct people who made inaccurate judgments about what the books said about him. People could say anything, and it would be "correct" because it could not be disproven.

Mangione, by virtue of his circumstances, was effectively cast as a fictional character. He was a text to be analyzed, a story to be consumed, a lesson to be learned. He was a celebrity of the one-way paradigm: He'd give you himself and ask nothing in return.

He was, in his extreme circumstances, emblematic of

the banal ones. Celebrity does more than turn people into fictions. It also turns people into images and texts—objects to be read, and read into, like evocative pieces of art. We take the information provided about them—what they might reveal of themselves by their clothes or their weight or their TikTok videos—and extrapolate, analyzing outward. And when anyone can become a celebrity, anyone can be read from afar. Anyone might be parsed ("MAGA-coded," "lib-coded," "queercoded") or dismissed ("sus"). Little wonder that some of the words that have caught on to describe exploitation, in digital environments, reflect the logic of text: marginalization. Erasure. People are "centered" in the story or they are removed from it. They control their own narrative or have a narrative imposed upon them.

The nineteenth-century poet Samuel Taylor Coleridge argued that fiction requires from its audience a "willing suspension of disbelief." The general notion today is so thoroughly accepted that it no longer seems like an idea at all. But Coleridge coined the term precisely because, in his time, the lines between fiction and reality were slightly different from what they are today. Fiction, in the nineteenth century, often doubled as moralism: You'd be lied to *as a lesson*, the logic went. Fiction might also imply supernatural forces and magic. But Coleridge wanted to write fictions that resonated precisely because they seemed realistic. He wanted his fantasies to be relatable. An audience that was willing to suspend disbelief would free him to weave fact and fiction, and allow him to assume that the audience would follow the story where it led. Coleridge called that two-way trust "poetic faith."

What the web has created, though, is a form of poetic cynicism. Real people, now, are providing the cultural fodder that fictional characters once did. We watch them. We read them. We analyze them. We bring the discerning eye of the art critic to the people who are there, and not there, on the screen. Online disinhibition is a personal delusion that risks becoming a collective condition.

"Half People"

In 2022, the Center for Countering Digital Hate (CCDH), a nonprofit that studies hate and misinformation in online spaces, analyzed more than 8,717 direct messages sent to five high-profile women on Instagram. One in fifteen of those messages, the research found, contained content that violated Instagram's own community guidelines. (The CCDH also reported that Instagram did not act on 90 percent of those abusive messages—even though their content was reported to the service's moderators.)

The online bullying that she received in response to a single comment she made about her sitcom, the actor

Constance Wu has said, got so bad that she considered suicide. Many other actors—almost all of them women, very many of them women of color—have reported similar types of abuse. And the verbal cruelties often manifest in the end as direct threats of violence. The actor Amber Heard was one of the women who participated in the CCDH's study. One of the messages she received—sent via voice note—said, "You, I don't like you, you are bad people. Die! Die! Die! Die! Die!"

But casual cruelty is not limited to social media's high-profile users. In 2022, a writer identifying herself as a thirty-year-old woman wrote to *Slate*'s sex-and-relationship-focused advice platform. In the process of seeking assistance, she offered testimony. "I don't know a lot of people where I live," she wrote, "and I had to stop using dating apps because people kept matching with me just to tell me that they thought I was ugly and I should kill myself. And I only have so much patience for that kind of treatment."

It's no coincidence that the age of social media has turned empathy into a cliche: a slogan, a plea, a concession. Target now sells T-shirts with BE KIND emblazoned across their chests, and the merch, itself, can read like a morality play gone wrong.

Kindness, in a functioning society, should operate as one of Émile Durkheim's social facts: something so obvious that it does not need to be articulated. Kindness, in our society, has become an argument.

"This Trial Is for Fun"

In the late spring of 2022, the actor Amber Heard and her ex-husband, the actor Johnny Depp, went to trial, each having sued the other for defamation. The proceedings centered around allegations of domestic violence that the former spouses had made against each other—physical assaults, verbal abuses, more. The testimony delivered during the trial repeated those accusations in blunt and often bleak detail. But despite the bleakness of the proceedings, or perhaps precisely because of them, the trial became a media sensation. The judge presiding over the case had allowed the trial not only to be recorded, but also to be broadcast in real time. The proceedings aired, then, like a TV show, on standard TV channels and on live-streamed websites: part soap opera, part true-crime procedural, part sporting event, part sitcom—a legal drama set within the wreckage of two real people's lives.

On social media—and on the comments sections that accompanied several of the web-based livestreams—many of the trial's viewers responded to the trial in the way it had been presented to them: as a running performance. Many of them decided, further, that the true genre of the performance was comedy. While cross-examining Depp, one of Heard's lawyers read a text message that the actor had sent about his former wife, expressing the hope that her "rotting corpse is decomposing in the fucking trunk of a Honda

Civic." At this a YouTube livestream erupted in digitized laughter. "LOL Honda Civic," someone replied.

On TikTok, people began repurposing footage of Heard's often tearful testimonies, stitching allegations of abuse into cheeky memes. (One of those people was Lance Bass, the member of the boy band *NSync: In a now-deleted post, he portrayed himself mugging to an audio clip of her testimony.) *Saturday Night Live* devoted its cold open to what it dubbed the "Depp v. Heard Cuckoo Trial." It featured the comedian Cecily Strong playing the trial's judge and saying at one point, "I'll allow it—because it does sound fun, and this trial is for fun."

The show was criticizing the way the trial's tragedies have been repackaged as spectacles. But its attempt at satire was complicated by the fact that the proceedings had already been made into comedy. "I've never been so scared in my life," Heard had testified about an alleged incident involving a broken bottle, a threat to "carve up" her face, and finally a sexual assault: "I didn't know if the bottle that he had inside me was broken," she said.

SNL, in its sketch, was doing roughly the same thing that internet audiences were, as they claimed the trial for their own amusement: The show gave the allegations a laugh track. And the audience howled in response. The cackle-face emojis, the memes, the likes, the lols, the lulz—these were the responses from the court of public opinion. Jokes make fun of things, in every sense. They can offer levity but also permission—in this case, to treat other people's pain as a punch line. The jokes doubled, in this trial, as motions to dismiss. Legal proceedings are, ostensibly, matters of fact: they revolve around evidence that is carefully culled. Their witnesses are vetted before they are sworn in. But jokes make the facts, effectively, irrelevant. Their only obligation is to make people laugh. In this way, and this way only, the trial succeeded—and the laughter was its own kind of verdict.

You might see in that laughter Durkheim's social current at work. You might also see how quickly the social current took on the force of a social fact. Individual people came together to observe these two famous actors turning into main characters of a different sort—and in short order the individuals became a collective, swept up in their shared endeavor. They mocked. They parsed. They liked one another's comments. They shared one another's memes. They read the trial and its participants with the license of the body-language expert. ("johnny depp's witnesses look honest and sincere while amber's looked like a bunch of liars," one trial-watcher declared in the chat that accompanied a trial livestream. Another: "No one on Amber's side is decent.") They read one another's reviews of the characters on the stage. They became fans of the reviewers themselves.

Courts traditionally do not allow cameras of any kind into their courtrooms—a practice that ensures that "courtroom artist" will remain a viable career path, yes, but one that remains precisely because cameras, like crowds, change the dynamics of things. The screens offered silent testimony in that courtroom, slowly transforming a human tragedy into a shared spectacle. Heard shared wincingly personal allegations in a setting that would typically be untelevised. In this one, though, the allegations became fodder—for memes, for commentary, for shared laughter. Enterprising creators profited from her pain, gaining likes and clout and new followings. Heard was the trial's designated villain, brought low so that other people could be elevated. She proved how possible it is, in the glare of the screen, to be centered and marginalized at the same time.

At least, though, she had not been cast as scenery. A few years ago, in Melbourne, Australia, a man approached a woman and asked her to hold a bouquet of flowers for him. She did. He walked away, leaving her with the bulky stems. He was a TikToker, it turned out, performing a "random act of kindness" as a bit. The video he posted of the interaction—without the woman's consent—reaped more than fifty million views. It also garnered effusive praise. (One comment: "Wow that was so beautiful I swear I would cry." Another: "My heart! That made her feel so good and it looks like she might have needed it.")

The woman on the receiving end of the "kindness," however, had a different view. The exchange had dehumanized her, she later said. It had insulted her. It had offended her. And it had made her feel, she said, precisely like what she had been, in that moment: "clickbait." The road to Panopticon, it turns out, is paved with good intentions. The woman and the TikToker were strangers. But they were strangers who were contending with the fact that, in the screens' frantic metropolis, we have yet to reach consensus on whether the people we encounter in public are people or props. "Kindness," he said. "Clickbait," she replied. You can tell a lot from the fact that they were talking about the same interaction.

People as props; people as extras; people as images; people as scenery: Anyone might find themselves cast out of their personhood. Celebrity will not exempt you. Ordinariness will not exempt you. Death, it turns out, will not exempt you. At the 2012 Coachella festival, Tupac Shakur, who had died more than fifteen years prior, reappeared as a digitally manipulated hologram, raising his arms triumphantly and then performing his posthumous 1996 single "Hail Mary." A version of Michael Jackson, who died in 2009, moonwalked across a stage in 2014. Amy Winehouse, Roy Orbison, and Elvis Presley have also been brought back as holograms.

Their resurrections—Lazarus by way of Lacan by way of lasers—are meant to be tributes to great artists: monuments scaled up to the digital age. They are also, and I believe this is the settled-upon academic term, creepy as hell. That is perhaps why, after rumors spread that the 2018 Super Bowl would feature a holographic Prince, fans reacted with horror and the idea was scrapped. (The artist himself had been asked in a 1998 interview whether he would avail himself of digital-editing capabilities that would allow him to play with long-dead performers. "Certainly not," he replied. "That's the most demonic thing imaginable.")

But the resurrections continue. *The Julia Child Challenge*, a recent Food Network competition show, uses extant footage of the beloved chef to create an Oz-like avatar that hovers over contestants as they recreate her dishes. She occupies, in the show, what the writer George W. S. Trow called "the context of no context": Disembodied and displaced, her words spliced, her movements controlled by unseen producers, she is simultaneously there and not there at all. The woman who, in life, was an avatar of lusty humanity—Child was warmth and joy and appetite—is flattened into an image. The tone of *The Julia Child Challenge* is reverent, but it indulges, in the end, in the soft bigotry of hagiography. Child was so beloved, the logic goes, that even death will bow to her fandom.

The personhood of holograms is likely not an area that many law students would consider specializing in. But the questions raised when celebrities are brought back from the dead implicate anyone who interacts in digital spaces. The rights of "digital replicas" and "synthetic performers"—representations of people both living and dead—are ever-more-standard considerations of the laws that govern intellectual property. They are matters of active debate within the Alliance of Motion Picture and Television Producers, the entertainment industry's biggest union. They are extensions of similar debates that the Screen Actors Guild and the Writers Guild of America have been engaged in as they attempt to apply legal language to the world-creating possibilities of artificial intelligence.

The debates are, in one way, specific to Hollywood. But, as with so much else, they will settle eventually on the rest of us. Screens make all of us "digital replicas." They make all of us "synthetic performers." The judicial precedents established in this moment, in this early age of AI, will resonate far into the future. The questions the courts are contending with, whether they concern holograms or AI replicas or some Frankensteined combination of the two, might seem to be distant from our own lives. They are not. This is one more way that the distance of the screen can render us short-sighted.

7
The Fans

In a 2006 scene from the NBC sitcom *The Office*, the beleaguered salesman Jim Halpert returns, after a brief time away, to the Scranton branch of the paper company Dunder Mifflin. He reunites with Kelly Kapoor, the bubbly customer-service representative. The two had developed over the years one of those relationships that long-term coworkers so often will: They are colleagues who are friendly but not quite colleagues who are friends. Reunited, they have the following exchange:

KELLY: Oh, my God, I have so much to tell you!
JIM: Really?
KELLY: *Yes.* Tom Cruise and Katie Holmes, they had a baby and they named it Suri, and then Brad Pitt and

Angelina Jolie, they had a baby, too, and they named it Shiloh. And both babies are *amazing*!
JIM: Great! What's new with *you*?
KELLY: [blank stare] I just told you.

The exchange is notably dated—and only in part because the "babies" in question, as of this writing, are old enough for college. But it also, as they say, *holds up*. Kelly, in her excitement to share her "news," is announcing a uniquely web-borne form of fandom. She is suggesting that her life and the celebrities' lives have merged so completely that they have become indistinguishable. Their news is her news. Their plot arcs are hers, too. Kelly's giddiness, in the scene, belies a profound and paradoxical kind of loneliness: She *feels* connected. She isn't.

Duane "The President" Johnson

There was a stretch of time, in 2021, when many believed that Texas might soon get a new leader: Governor Matthew McConaughey. Media outlets speculated that the actor might challenge the incumbent, Greg Abbott, for the role. They wrote op-eds about the possibility. They conducted polls. They joked that McConaughey's campaign, like his catchphrase, would be "alright, alright, alright."

The actor himself was relatively silent about his plans (or, more accurately, his lack of them—he did not, in the end, stage a run). But no matter: Political campaigns, as a rule, are more enjoyable in theory than in action. "Texas may not be ready for a philosopher king as a candidate, much less governor," *The New York Times* wrote, "but it sure would be fun to watch Mr. McConaughey debate Mr. Abbott."

McConaughey's conscription put him in good company. Oprah Winfrey is a regular subject of "should-be-president" chatter. Dwayne Johnson—the wrestler-turned-actor better known as "The Rock"—revealed that he had been approached about a run by party operatives armed with "deep-dive" research and polling data. The drafting efforts for the most part are testaments to the draftees: evidence that they have achieved a form of fame so transcendent that it has also become bipartisan. But they are also testaments to the times.

The notion that an actor would be fit to run a government—of a state or of a nation—might be on its face somewhat ridiculous. We don't typically expect that surgeons will make good fighter pilots or poets good CEOs. But Governor McConaughey resonated as an idea precisely because Americans no longer treat "politician" as a job. It is a role. It is a performance. And it is one that eschews competence—pities it, really—and rewards charisma.

George W. S. Trow located the origins of that assumption to the 1950s, when a decorated general, in the name of relatability, shed his stripes. "In the phrase 'I Like Ike,' the power shifted," Trow observed. "It shifted from General Eisenhower to someone called Ike, who embodied certain aspects of General Eisenhower and certain aspects of affection for General Eisenhower. Then it shifted again. From 'Ike,' you could see certain aspects of General Eisenhower. From 'like,' all you could see was other Americans engaged in a process resembling the processes of intimacy."

But there are many such moments you could see as touch points in the transition from politics to political theater: the presidency of Ronald Reagan (the former actor won a second term in a field that included another celebrity: the former astronaut John Glenn); the tentative, awkward appearance that Richard Nixon made on the comedy revue *Laugh-In* ("sock it to meeeee"); the efforts made by presidential candidates, from Gerald Ford to Kamala Harris, to prove their ability to lead the most powerful nation in the world through appearances on *Saturday Night Live*.

You could also point to the moment when, in June of 1992, the governor of Arkansas played his saxophone on *The Arsenio Hall Show*. Bill Clinton, donning sunglasses, played Elvis Presley's "Heartbreak Hotel" as the show's live audience applauded and cheered. When Clinton finished the song, their standing ovation went on for so long that Hall had to motion for them to sit down again. Clinton's path to the Democratic presidential nomination had been stymied by scandals that had included allegations of sexual misconduct; the late-night performance, though, had a cleansing effect. Delivering that wordless performance on the late-night stage—intuiting that the talent show, not the interview, was the crucial component of the pageant—the young candidate rewrote the rules of political engagement. He also, in some sense, reversed them.

Ronald Reagan had been, technically, the first celebrity president. He was a celebrity, though, who had transformed himself into a leader. While he leveraged his Hollywood connections in his bids to be sent to Washington, he sold himself, fundamentally, as a conventional politician. Clinton, on the contrary, was a conventional politician—a governor, a lawyer, a policy wonk—who remade himself, day by day, as a star. He sold himself as a leader by presenting himself as a performer. He was taking the "beer question" that would become ubiquitous in polling—"Which candidate would you rather have a beer with?"—to its logical conclusion: If Americans seek relatability in their leaders, the most relatable thing of all is the thrill of a good show. Winning votes, Clinton's campaign assumed, amounted to winning fans—and fans are won not through facts but through feelings. In that,

Clinton was prescient: Politics is, evermore, a matter of abstract devotion.

"When politicians and bureaucrats judge themselves only by their ability to seduce the public," the writer Christopher Lasch observed, "they lose objective standards by which to measure themselves—or to measure the success or failure of their policies." He wrote that in 1979. Today's politicians do not merely measure themselves by their ability to charm the public; they define themselves by it. But the logic of fandom and the illogic of it are no longer merely components of the relationship between leaders and their constituents. Very often, now, leaders come to office assuming that fandom, in politics, is all there is—that performing politics and actually engaging in them are, in the end, the same thing.

The sentiment is ever more common—not just in Hollywood but in Americans' political habits. This is what happens when Americans become divided, as Stephen Colbert put it, between "those who think with their head and those who know with their heart." The logic of fandom offers people permission to engage with political events—and with politicians themselves—in terms that are purely visceral: vibes, all the way down. "I can't really put my finger on it, but something just doesn't feel right," a Wisconsin woman told a pollster about Donald Trump's (thoroughly unsupported) claims of election fraud. A man from Pennsylvania said of Joe Biden's win, "Something about it just didn't seem right." Another man, from Arizona: "It didn't smell right."

Like and Subscribe

In 2022, a new podcast joined the fray. *The Langley Files* offers a mix of oral histories and interviews conducted by its two hosts, Dee and Walter. The show is consistently earnest, occasionally interesting, and for the most part thoroughly conventional—save for the fact that it is a production of the CIA.

The Langley Files, its marketing literature promises, allows listeners to "step beyond the Hollywood scripts" to learn about the agency as it was and as it is today. What the series finally offers, though, is a reminder of how inescapable those scripts have become. Fandom's currency is so strong that even the clandestine service felt compelled to trade in it. Spies have long assumed new identities to keep their real ones hidden; only in this age would they adopt the guise of the content creator.

"Corporations are people, my friend," Mitt Romney declared in 2011. The line was roundly mocked at the time (a representative tweet: "I FEEL TERRIBLE: Exxon had a birthday, and I didn't get him anything but subsidies"). But it was merely ahead of its time. The company and the performer, the influencer and the audience, the person and the statistic—they are merging, now, along with so much else. "Corporate personhood" has transitioned from a legal debate into a cultural banality. The "personal brand," having long ago lost its novelty, is well on its way to shedding the last of its smarm.

Along the way, though, the personal brand is making the leap from people to information. "For Gen Z, TikTok Is the New Search Engine," *The New York Times* announced in 2022, suggesting how the benefits of the "personal brand" could make their way into Americans' broader informational ecosystem. The same year, *BuzzFeed News* reported that social media had given the old commercial line—"I'm not a doctor, but I play one on TV"—a surreal new twist. Doctors who had gained large followings on TikTok and Instagram, the site reported, had found a way to navigate the "red tape" of the medical system: through recreating themselves as cartoons who share medical advice.

The article is thoroughly reported and explores all the related ideas you'd hope it would: the market opened for these doctors by an unwieldy healthcare system; medical privacy; medical expertise; medical malpractice. The story also notes that, despite its reporting efforts, it was unable to determine the current qualifications of the doctors who were promising to "revolutionize medicine."

Instead, "being popular on social media appears to be one of the most important credentials."

Doctors-as-influencers can be a way to bridge the gaps between healthcare as a system and healthcare as a right. They can help people to feel agency over their health. They can make the sometimes awkward business of having a body seem a little more manageable.

When health is at stake, though, the comedy can quickly become the stuff of tragedy. Some of the people who died during the Covid-19 pandemic lost their lives not strictly because of the virus but because of widely shared misinformation about how the virus might be stopped. But the lines, on TikTok in particular, blur. Medical professionals are educated, licensed, required to keep updating their expertise. Medical influencers are required only to acquire large followings, however they see fit to do it. Their crowds are their credential.

This can, in one way, democratize medicine, allowing people-turned-patients to turn their own health into a script for their own writing. But there are scripts and there are scripts. And the collision of doctor and influencer can also lead to confusion. It can lead to absurdity. It can require warnings like the one that a medical toxicologist—an actual doctor—shared with *BuzzFeed*: "If your child is in respiratory distress in the middle of the night, having a surgeon famous for dancing on TikTok text you might not be that useful."

This is a category collapse of a sweeping kind. And the collapse has required TikTok, as a platform, to contend with many of the same descriptive challenges that Facebook, Twitter, and similar services have: Is it, fundamentally, a platform, or a publisher? Is it providing, to its legions of users, entertainment or information?

TikTok, like its predecessors, has at this point tried to have things both ways. Its executives have insisted that they recognize the "responsibility" that comes with the large audiences that tune in to their shows; they have further stressed that they take the responsibility seriously. But they have also insisted that TikTok offers entertainment, fundamentally—which also suggests that the responsibility they acknowledge, entertainment being what it is, only goes so far.

Historical Fan Fiction

The Greatest Showman, 2017's star-spangled retelling of the life of P. T. Barnum, is a film fit for its subject. Starring Hugh Jackman as the pioneering entertainer, the movie is in one sense a standard-issue biopic. It offers insight into the man behind the myth—Barnum the businessman, the family man, the visionary stymied by other people's shortsightedness—but with such naked admiration that the myth emerges stronger than ever. *The Greatest*

Showman is *based on a true story*, with an added twist: It is a musical. History, here, comes to life so exuberantly that it regularly bursts into song.

The showman, remade as a show: It might have been a lyrical way to consider both who Barnum was and the world he helped to create. But *The Greatest Showman* does more than remake Barnum's life as a musical. It remakes Barnum, the historical person, as a hero fit for modernity: an entrepreneur, an empath, a champion of the oppressed.

In life, Barnum was the person most directly responsible for adding "freak show" to the American vernacular. He "displayed" Joice Heth, a woman who had been born into slavery and whom Barnum purchased—*purchased*—in 1835, promising his patrons that she was (1) 161 years old and (2) the former nanny of George Washington. When Heth died, Barnum found a way to keep her as a box-office draw: He arranged for her to be autopsied before a crowd of 1,500 people. When the doctor performing the surgery pronounced that Heth had been around 80 years old when she died, Barnum claimed that the body on the table was not Heth at all. The real Heth, he insisted, was so hardy despite her 161 years that she had embarked on a tour to Europe.

That was P. T. Barnum: exploiting, denigrating, and then offering people the opportunity to be lied to by the flashiest liar of them all. (Later in life, as such performers are wont to do, Barnum became a politician. With the unique sanctimony of the con man, he championed a Connecticut state law that would imprison couples who used contraception. The law passed; its extremism became the basis for *Griswold v. Connecticut*, the landmark Supreme Court case.)

You would learn none of that, though, from *The Greatest Showman*. The film is a biopic with very little *bio* and quite a lot of *pic*. And in one way, of course it is: It's a glitzy musical. The movie's promise, like its hero's, is that "reality" is much more compelling when it comes with scare quotes. But that's also why I mention it. *The Greatest Showman* is so deeply ahistorical that it's probably more accurate to call the movie an antihistory. But it is also, when it chooses to be . . . quite historically accurate? The real Barnum did do much of what the film claims: Attempting to dehumbug his act, he brought the singer Jenny Lind—the "Swedish Nightingale"—on tour in the States. A turning point in his life, as in the film, was a fire that razed one of his theaters. At some point he might even have had a colleague who looked like Zac Efron.

The unsteady terms of it all—the sleek interplay of the real history with the imagined past—might not be a widespread problem in normal times. But these are not normal times. These are times when the facts of the world are becoming ever more Barnumized. It is not a coincidence that the moment that brought "alternative facts" into the American vernacular is the same moment that regularly finds people outraged at the reminder that George Washington enslaved people, or that the man who "discovered" America was also an agent of genocide. Every day, indignant Americans express outrage not at the injustices of the past but at those who acknowledge them at all. Washington *was a hero*, they shout, and to speak of the fuller picture is to disrespect the man and, by extension, the country. Soon, the logic transfers to any variable that might complicate the delicate equation, and before long history ceases to be history at all and instead becomes a ceaseless act of fan fiction.

In late January of 2025, President Donald Trump issued an executive order calling for a restoration of "patriotic education" in the nation's schools, from kindergartens to twelfth-grade classes. The order was arguing, effectively, that American students should learn the fan-fictionalized version of the past. It was making the case for mythology—warm, inspiring, comforting, easy—over history.

History, as Lauren Berlant argued, hurts. But when we cede its genre—when we insist that the American story is a great epic, its stars cinematic, its plot cathartic—we also cede, in a very real sense, ourselves. We subject the past itself to the demands of American exceptionalism. And the facts that prove uncomfortable, in that environment, easily begin to seem fungible. Their tone is wrong. Their vibes are off. They serve neither the A story nor the B. But edit history cannily enough, and its genre becomes epic once more. Remaster the American story in Technicolor, and you'll fulfill the epic's only obligation: to be a spectacular show.

8
The Haters

The boy who cried wolf cried because he was bored. Tending to sheep can be lonely, dull work, as Aesop observes at the start of his fable—so the young shepherd, lacking ready-made fun, decided to make some of his own. He ran through the nearby village, shouting warnings of a wolf in the pasture; when the villagers came to defend the flock, they found nothing but the laughing boy. A few days later, he played the same prank. They came again. He laughed again. And then . . . well, you know the rest. A wolf prowled; the boy yelled; nobody helped. The wolf had its feast, and Aesop had his lesson: "Liars are not believed," he warned, "even when they speak the truth."

The boy is the star of this terse little drama; the villagers are extras in the show. But it's the villagers, today, who propel "The Boy and the Wolf" into relatable modernity. While the boy learns precisely nothing over the course of the tale, the villagers hurtle along Aesop's version of a character arc—trusting, tricked, chastened—leaving the story different from the people they were before. They begin as a community, ready to help a member in need. They end as individuals, wary and torn.

The villagers, through these sparse, ancient lines, seem to sigh from across the centuries. Squint, just a bit, and their tale becomes a lesson about life in our own village, with our own lurking threats, as we find ourselves caught, again and again, in other people's fun. Through them, Aesop's story becomes a parable about uncertainty and entropy, about the chaos that sets in when the cries of wolf we keep hearing might be real news or fake news or a bored boy's thoughtless joke. It becomes a tale about reality and "reality," tricks and trolls, fact and fiction—and the impossibility, soon enough, of telling the difference.

The boy is not a partisan, necessarily. He's just a kid who got bored. But his antics provide a tidy lesson in how partisanship can operate. Partisanship can allow people to look at things without seeing them. It can turn fandom against itself, curdling devotion for one's own side into antipathy for the other. It can weaponize the familiar language of the show. "Crisis actors" are favorite scapegoats of conspiracy theorists—and of those who, more broadly, find it inconvenient to admit to truths that are plain. "Fake news" is a common response not to assessments of a news story's factuality but instead to assessments of its palatability. It has been joined by airier accusations of fictionhood: "I'm convinced it's a bit," or "this must be a joke," or "this must be performance art."

At its most extreme, "it's all a performance" can enable horrifying acts of cruelty, as when Alex Jones—for years—claimed that the murdered children of Sandy Hook were somehow, in fact, crisis actors.

In 1990, GOPAC, a political action group headed by the representative Newt Gingrich, mailed a pamphlet to Republican candidates who were running in state elections across the country. The mailing, an adjunct to cassette tapes that the group sent to help prepare candidates for their runs, was written in cheerful, conversational tones. "As we mail tapes to candidates," it read, "and use them in training sessions across the country, we hear a plaintive plea: 'I wish I could speak like Newt.' That takes years of practice. But we believe that you can have a significant impact on your campaign if we help a little."

The rest of the document doubled as a dictionary. It provided 133 words that electoral hopefuls might use as they sold themselves to their potential constituents—words that would elevate themselves (*family, freedom, pride*) and vilify their competitors (*decay, betray, corruption, pathetic, incompetent, shame, disgrace, punish, lie*). The memo's upbeat introduction was belied by its title: "Language: A Key Mechanism of Control." Many in the media, nodding to the pamphlet's dark resonance with the language used in Orwell's *1984*, came to know it as "Newtspeak."

The 1990s were years that found many politicians paying new attention to language—and translating some of the truisms of postmodern analysis (the power of "framing," of narratives, of metanarratives) into the argots of everyday politics. This was the era of *Wag the Dog*. It was the era of Peak Limbaugh. It was the era when political optics were encroaching into political realities. It was a moment, in culture as well as politics, of widespread cynicism. The things Walter Lippmann had identified as essential to society and to democracy—the pictures people hold of one another, the assumptions they make of the world—were becoming at once harder and more slippery. "Spin" was both an accusation and an expectation of political discourse.

The GOPAC pamphlet crystallized the shift. It turned spin, that longstanding mode of political persuasion, into a plot twist. Words shape the world even as they reflect it: Making their way into people's language, they can also change people's minds. And the memo's recommended vocabulary—additional terms included *traitors, selfish, insensitive, steal, destructive,* and *sick*—treated political discourse, in one way, as deeply personal. Democrats, in this dark new dictionary, were not merely one's opponents;

they were one's enemies. The lexicon recast American politics not as an ongoing debate among equals but as an epic battle between right (*crusade, control, truth, moral, courage*) and wrong. The differences between the two sides were not merely political, the words suggested; they were moral.

But the words' recommended usage, in the memo, was also strikingly impersonal. The memo classifies *traitors, sick, pathetic,* and their ilk merely as "contrasting words," and suggests that its readers should feel no compunction in using them. "Sometimes we are hesitant to use contrast," it says. "Remember that creating a difference helps you. These are powerful words that can create a clear and easily understood contrast. Apply these to the opponent, their record, proposals, and their party." The memo's closing page offers a range of venues for that application, among them speeches, direct mail, brochures, flyers, newspaper ads, and door-to-door campaigning—basically, "anywhere you need to talk about your campaign, your vision for a better community, and the new ideas and approaches you represent!"

The vocabulary ("read them," the memo urged, and "memorize as many as possible") was at once extreme in its rhetoric and all-purpose in its usage. It was modular: The words could be combined however the candidate saw fit. They were not directly descriptive, but instead vaguely evocative. And they could be applied to anyone, and to anything. As a matter of political duty, indeed, they should be. One may be "hesitant to use contrast," as the memo put it, but one must get over it: "Contrast" is too powerful a weapon to let go unused. "Contrast" obviates the need for evidence or explanation; when you call your opponent a traitor, you have taken things out of the realm of reason. You have served your fans by giving them enemies.

"Language: A Key Mechanism of Control," because of that, was a tipping point. It set a "new course," *The Atlanta Journal-Constitution*, Gingrich's hometown paper, argued in 2016. The pamphlet, in its cheerful assurances that aspiring politicians could learn to "speak like Newt," also reads like a skeleton key—a word-by-word explanation of our own political moment. GOPAC's 1990 release of the memo roughly coincided with the release of the World Wide Web. And the memo, in framing politics as both extremely personal and extremely impersonal, anticipated the political environment that would thrive on the interactive screens. The memo, in its own way, was applying that defining question of digital life to political discourse. "Is that person . . . real?" it asked, effectively, about Republicans' political opponents. "Is that person *a person*, or something else?"

Once those questions were asked, the "contrasts" could be deployed for full effect. The world would be reduced to "real Americans" and those who, by implication, were not. It would be narrowed to teams and tribes, matters of "us" and "them." *Traitors, betray, pathetic, lie*—the words, today, have retained their extremism. But they have lost their novelty. They are ubiquitous. They are banal. They bring an atmospheric ease to Huxley's insight about propaganda: that its purpose, in the end, is to make one group of people forget that another is human.

Martial Laws

"We're in a great time of spiritual warfare," the political strategist Steve Bannon said in 2022. "It's the forces of good and it's the forces . . . of evil." With the observation, he was taking an accident of language—the tumult of democratic engagement, described as a "campaign"—and transforming it into an ethic. He was suggesting that "culture war," that all-purpose explanation for Americans' political ills, might be best understood not as a metaphor but as a mandate.

The populist politician Pat Buchanan is often credited with unleashing the language of the "culture war" onto the American public. ("There is a religious war going on in this country," he informed his audience at the 1992 Republican National Convention. "It is a cultural war, as critical to the kind of nation we shall be as the Cold War itself.") But the credit for the concept—or, perhaps, the blame—should more properly go to James Davison Hunter, the sociologist who popularized the term in his 1991 book *Culture Wars: The Struggle to Define America*.

Hunter, who studies the intersections of religion and culture, borrowed the phrase from the German Kulturkampf ("culture struggle"), a term that had previously referred to the clashes between the German empire under Otto von Bismarck and the Catholic church. While the phrase had made occasional appearances in the context of American politics, Hunter repurposed it in a way that stuck, mainly as a description of what he had been observing at the tail end of the Reagan presidency: a cultural polarization that was playing out as a political one.

Culture, Hunter observed in a 2018 interview with *The Wall Street Journal*, is "about systems of meaning that help make sense of the world, why things are good, true and beautiful, or why things are not. Why things are right and wrong." Culture also "provides the moral foundation of a political order"—even as the state "becomes the patron of a certain vision of the world." One consequence of the culture war in Hunter's conception can be a feeling of ambient siege: When the faction you consider to be your adversary is in power, the *Journal* put it, "it can feel as if you are under hostile occupation." The war metaphor further comes into play because "the state is the institution that holds the reins of legitimate violence," Hunter said, "and this is one of the reasons why our disputes tend to be litigated more than they are actually debated."

War is horror that can function, in practice, as permission. The demands of the battlefield—the chaos, the danger, the establishment of allies and enemies—rationalize behavior that would, in other settings, be unthinkable. Newtspeak, that louche new lexicon, attempted to apply the exceptionalism of martial campaigns to the practice of political ones. If your side is the right side, the logic goes, you can do—indeed, you are obligated to do—whatever you must to ensure your side's victory. And then you must ensure that the winning never ends.

"They are vicious, horrible people. . . . They are horrible people," Donald Trump said in 2020, applying the permissions of warfare to Democrats, personally and generally, as a group. In the fall of 2022, the Georgia representative Marjorie Taylor Greene told a group of young people in Texas that her Democratic colleagues are "night creatures, like witches and vampires and ghouls."

The absurdity of such claims (night creatures! ghouls!) belies their impact. They were, after all, Huxley-level assertions. Greene was not merely insulting her colleagues; she was rejecting their humanity. In 2022, the Public Religion Research Institute conducted research into the hold that QAnon, the community and the conspiracy theory, had taken over Americans at large. In one question, the researchers asked survey respondents whether they shared the QAnon belief that "the government, media, and financial worlds are controlled by Satan-worshiping pedophiles." Nearly a fifth of the respondents—16 percent—said they did.

This is what happens when people become fictions to one another. Logic leaves. Anything seems possible. Scholars talk about epistemic closure, one of the cognitive biases that inform conspiracy theories: In political contexts, the bias refers to beliefs so firmly held that they will permit no evidence to contradict them. "There are no coincidences," one of QAnon's slogans insists, and that is epistemic closure at work. QAnon, relatedly, adheres to the gospels of entertainment: It harnesses the forensic satisfactions of the true-crime procedural. It assumes that its prophecies will play out on live TV. It trusts, above all, in Q, the anonymous showrunner who is writing and directing and producing reality—and, every once in a while, dropping tantalizing clues about what might happen in America's coming seasons.

But this particular form of trust can close in on itself, betraying the believers and making them vulnerable, instead, to misinformation, to propaganda, to hatred. Villains may be necessary features of movies and shows: Good plots require conflict, and villains are discord incarnate. But when the enemies are real people—and when they constitute approximately half the country—the discord becomes untenable. Enemies everywhere is a great approach to drama. It is a much less great approach to life.

How to Look Away

Popular culture has always served as education by other means. This is part of its power. It is also part of its anxiety. (The encroachments of "edutainment" have made for longstanding worries among cultural critics.) The problem, though, as it so often will be, is a matter of scale. The works of blurred reality proliferate, financed by wealthy conglomerates and the demands of streams that will never be filled. The reliable information, meanwhile, suffers. Local news is dying away. Regional and national news outlets are, as a whole, shells of what they once were. Nothing has come in their place.

When big news breaks now, many people will offer up verbal shrugs in response: "I'll wait to care until it's a six-part HBO miniseries." This is a joke, sort of. But it's a revealing one. After a while, the fictionalized shows come to take the place of the straight-ahead facts. The suspension of disbelief becomes, more simply, the suspension of attention itself.

Deference to "the later," in the scholar Lauren Berlant's terms, allows people to "suspend questions about the cruelty of the now." News becomes entertainment. Entertainment becomes news. People get their sense of the world's realities through pieces of semifiction whose success has very little to do with "is it true?" and everything to do with "is it entertaining?"

The story, as a term, makes no distinction between fact and fiction. Recent years have brought new acuity to long-running discussions—in literature, in philosophy, in life—about the possibilities, and the limits, of narrative as a form. "Story lulls," the critic Parul Sehgal wrote in *The New Yorker* in 2023. "It encourages us to overlook the fact that it is, first, an act of selection. Details are amplified or muted. Apparent irrelevancies are integrated or pruned. Each decision is an argument, each argument an imposition of meaning, each imposition an exercise of power."

This is absolutely true. Equally true, though, is that narrative is, in a very direct sense, all we have. To interrogate it very quickly becomes an exercise in tautology: We need the brute machinery of the story even to argue about the machine's obvious flaws. "We have, each of us, a life-story, an inner narrative—whose continuity, whose sense, *is* our lives," the neurologist Oliver Sacks wrote. "It might be said that each of us constructs and lives a 'narrative,' and that this narrative *is* us, our identities."

We expect our news, after a while, to twist itself into the shape of a story. *The medium is the message* also applies to narrative structures of TV and movies and video games: We expect the world to hew to the dynamics of the story, with tidy plots and cathartic conclusions.

Many of the dystopias imagined in the twentieth century—fictions molded by the realities of totalitarianism—treat television as a means of captivity. Big Brother monitors you, always, from the screen. The people of Brave New World lose themselves to their fantasies and their fun. But a more recent dystopia presents another possibility: the television that begets cruelty.

The Hunger Games, that distinctly twenty-first-century dystopia, flips the older scripts. In Suzanne Collins's political universe, TV captivates people in both senses of the term. Its programming lulls people into complacency. But the programming—in particular, the reality competition that pits twenty-four young people against each other in a gruesome fight to the death—also inures them to cruelty. It drives them to bloodlust. It makes them complicit in the violence of the regime.

Under its influence, citizens of the Capitol become capable of watching children murder one another and categorizing the whole thing as a good show. People's vision gets skewed. Their imaginations harden. They look at a bunch of kids, with their arrows and bombs and spears, and see not kids at all, but characters who perform for their fun. The contestants win fandoms. Their lives depend on their facility with weapons, yes, but also on their ability to charm the audience. Their pain becomes melodrama. Their deaths become plot twists.

The Triumph of the *Could*

In 2023, the Texas senator Ted Cruz posted an image that purported to be a screencap of an *Atlantic* magazine "cover story." The article was headlined "The Evolution of White Supremacy." Its subheadline explained the content of the alleged story: "In Dearborn Michigan, Muslim parents who oppose teaching pornography to children become the new face of the far right."

The image was entirely fake—a fact that approximately twenty elements of the image made immediately clear. Let's, for a moment, leave aside the fact that a United States senator feels the need to clout-chase by mocking large swaths of his fellow Americans on Twitter. ("the Left is beyond parody," @tedcruz wrote as the caption to the photo.) And let's further table the fact that the same senator, or that senator's staffer, failed to recognize any of the clues—among them the nonsensical phrase "teaching pornography to children"—that the image they were sending out to 5.3 million followers was a lie. Let's focus, instead, on what @tedcruz did after spreading the image that was, in the most literal way possible, fake news.

"Didn't know it was fake," the account tweeted. "You guys are so insane, it could easily have been real."

And so: Here was a high-profile representative of "the greatest deliberative body on the planet"—a lawyer who received his legal training from Harvard—resting his case on an argument that would get him laughed out of any courtroom: "But it *could* be true."

Could is easy. *Could* is irrefutable. *Could*, to paraphrase Hannah Arendt, makes *everything possible, and nothing true*. It is a mainstay, as such, in the scripts of our political theater. "Opinion: January 6 hearings could be a real-life summer blockbuster," read a headline in the summer of 2022—the unstated corollary being that if the hearings in question failed at the box office, they failed at their purpose. The events of January 6 had no real need for embellishment. They were a plot twist—a made-in-the-USA coup attempt—and set piece. Never has a national emergency arrived on the scene so thoroughly camera-ready.

The hearings remade them as a legal drama. The congressional committee investigating the attack hired a veteran TV executive to "produce" its findings for public consumption. It summoned witnesses who were well-spoken and, in many cases, notably telegenic. It made a point of transforming that day's chaos into a comprehensive plot. The production was so successful that *The New York Times* included the hearings on its list of 2022's best TV shows.

And yet, for some, the tedium was the message. "Lol no one is watching this," the account of the GOP House Judiciary Committee tweeted as the hearings were airing. The counselors, as they registered their objections, were testing out a bold new theory of jurisprudence: When the jury expects every fact to be fun, "it's boring" becomes a legal argument.

And it can be effective, as an argument, because it declines to do much arguing at all. Like "performative" and "inauthentic" and "it could be true," "it's boring" is an accusation that can't be answered. "What we will witness today is a televised theatrical performance staged by the Democrats," the representative Devin Nunes told reporters in 2020, at the outset of the first impeachment hearings of Donald Trump. As Nunes brought the insults of Instagram ("*so performative*") to the political theater, he helped to explain why the theater itself has become ever more inescapable. We can dismiss anything on the grounds that it doubles as a show.

The spin wasn't merely a talking point among the president's allies. The impeachment's lack of melodrama was also a common complaint among the journalists covering the event. "Unlike the best reality TV shows—not to mention the Trump presidency itself—fireworks and explosive moments were scarce," Reuters lamented soon after the hearings began. The first witnesses called in the trial, NBC News reported, had provided clarifying testimony—but "lacked the pizzazz necessary to capture public attention."

The news media were not alone in their disappointment about the viewing that results when the Constitution gets optioned for a TV show. *Saturday Night Live*, too, gave the hearings a thorough pan. The whole thing had consisted of "two weeks of dry debate and posturing," it lamented. The show's cold open promised to correct Congress's mistake, recreating "the trial you wish had happened." This proceeding did have pizzazz: It was a wacky musical comedy.

Our discomfort with boredom—our assumption that facts have an obligation to be fun—follows the familiar path: It inflicts itself on things, and then it inflicts itself on people. Invoked as performance, boredom can also send messages about who, and what, is worth one's attention—and about who, and what, is not. It can be a totalizing form of spin, and one that operates far beyond the arena of partisan politics. In 2020, during the criminal trial that found the film producer Harvey Weinstein guilty of sexual assault, one of his more than eighty accusers, Jessica Mann, testified against him. Mann told the court about Weinstein's "unpredictable anger." She described an episode of assault. As she spoke, she wept. Weinstein, meanwhile, seemed to be taking a nap.

Boredom is a posture that claims not to care; boredom, though, invoked as a show, cares deeply. It simply turns the care inward, refusing to see, to hear, to react. Weinstein sleeping as Mann spoke was an insult to her, yes; it was also a tidy portrait of disordered empathy.

Shipping Humans

Vision, whether literal or figurative, is a moral force. The way one sees people, or fails to, determines most everything else. It can turn empathy into actions: To "be seen" can be to be understood, aided, tended to. To see someone can be—and, then, can lead to—an act of care. But the opposite can be true, as well.

In September of 2022, agents working for the Florida governor Ron DeSantis convinced a group of people who had come to Florida seeking asylum to board airplanes. The convincing was done, reportedly, with the help of lies: The agents promised the asylum-seekers that housing, financial assistance, and employment would be waiting for them when they landed. Instead, the planes flew to Martha's Vineyard, where there was nothing waiting for the confused travelers except a group of equally confused locals. But, rallying, they gave the travelers food and shelter. Immigration lawyers came to help. Journalists obtained copies of the brochures handed out to the asylum-seekers, thus publicizing the series of false promises through which human beings were turned into props.

One might note, first of all, that the orchestrators of the stunt seemed unable to imagine that the residents of Martha's Vineyard would respond to the manufactured crisis by treating those they dismiss as "illegals" as people at all. One might also note that, confronted with their kindness, DeSantis spun the thing in precisely the direction you'd expect. To argue that the asylum-seekers were exploited and mistreated was to engage, DeSantis said, in "virtue-signaling."

The ploy was also a classic pseudo-event. It began as a wish and a joke: "Shipping migrants," for a while, was a regular topic of conversation on the Fox News morning show *Fox & Friends*, as hosts filled their air with fantasies about sending "illegals" to places that would be disruptive to Democrats, thereby punishing both. The idea was repeated so steadily that, as so often happens, the joke somehow became the plan, and then the plan became the reality, and then the asylum-seekers, desperate and misled, were routed like shipments of Amazon Prime to a place selected for the fact that Barack Obama vacationed there.

And the directors of the whole thing, rather than questioning the premise of their show, instead promised more performances. Ted Cruz—whose father, as it happens, came to the US as an asylum-seeker—announced his intention to send another group of migrants to Joe Biden's vacation spot. ("Rehoboth Beach next," the senator said.) The Texas governor Greg Abbott announced that he would be busing migrants out of the state—and that the first location for drop-off would be "the steps of the United States Capitol." The National Republican Senatorial Committee, not to be outdone, brought audience participation to the show: A fundraising email asked recipients to share their ideas about where the deliveries should be sent next.

Lolling Alone

In September of 2024, as the American presidential campaign was escalating to its fever pitch, Secret Service agents apprehended a man near Donald Trump's golf course in Palm Beach. He was carrying an AK-47-style gun. He was attempting, the FBI soon concluded, to assassinate the former president—just months after another man had made his own failed attempt on Trump's life at a rally in Pennsylvania. The efforts, to some, were evidence of the bleak state of American politics: assassination attempts joining war as "politics by other means." To Elon Musk, though, the gunmen's efforts were an opportunity. "And no one is even trying to assassinate Biden/Kamala," the billionaire wrote on X, the platform he'd bought in 2022 and renamed from Twitter in 2023. He punctuated the observation with a thinking-face emoji.

There was a time when irony was alleged to have died—a stretch of time, after the attacks of September 11, 2001, when humor itself seemed wrong. "None of us are feeling funny," an *Onion* editor told *The Baltimore Sun* on September 14. A Comedy Central spokesperson, speaking to the Associated Press, made a similar claim. "There's going to be a seismic change," Graydon Carter, then the editor of *Vanity Fair* magazine, told the site inside.com. "I think it's the end of the age of irony."

The reports of irony's death, it would turn out, were greatly exaggerated. Satire, now, saturates political discourse. Irony, with its teasing ambiguities and its plausible deniabilities, is everywhere. Was Musk joking when he dragged Trump's political opponents into the news of his attempted assassination? Was he suggesting that assassination attempts were approval ratings in disguise? Was he implying that Biden and Harris weren't worth the trouble of assassination? Was he daring people to prove him wrong?

There was no way to know for sure. Irony has not merely survived into the current era; it defines it. Every day, we play the part of Aesop's villagers in a sequel nobody asked for: What if the boy who cried wolf had also cried just kidding? Like the words of the GOPAC memo twisted into a sensibility, irony empowers the people who use it, and addles most everyone else. It might mean anything. It might mean nothing. It is, because of that, uniquely effective as propaganda. Earlier regimes found their power in the "triumph of the will." Our era relies on the triumph of the *could*.

Words that wink, since they require constant acts of microtranslation, can be exhausting. The scholar Dannagal Goldthwaite Young, analyzing fMRI studies that illustrate how the brain processes jokes, argues that humor can impose a cognitive tax. Jokes, for all their delights, ask more of their audiences than other forms of discourse do: They require more split-second parsing, more energy, more work. Irony, with its inherent instability, makes similar demands. When the world is always winking—or when it always might be—making sense of it can become exhausting.

Humor is an age-old political tradition: *Common Sense*,

the pamphlet that persuaded many Americans to become revolutionaries, was powerful in part because it was often quite funny. But irony, summoned into politics, can be divisive rather than galvanizing. Nazis both past and present, insulating themselves with irony, have hidden their hatreds in plain sight. Some of them, as the saying goes, have "said the quiet part out loud." Many others have shouted the quiet part, and then added a quiet "just kidding."

Humor can be self-rationalizing, and therefore particularly powerful when its logic comes for our politics. Donald Trump once gave a speech in the rain and then bragged about the sun shining down on his performance. His bravado was propaganda in its most basic and recognizable form—overt, insistent, blunt. It did what propaganda typically will, imposing its preferred reality onto the one that actually exists. But the lie was also so casual, so basic, so fundamentally absurd—even the heavens, Trump says, will do his bidding—that it barely registered as propaganda at all.

Trump came of age as a public figure in the 1980s, when partisanship was acquiring its martial edge and cynicism was expanding as cultural currency. The times shaped him then; they still do. Trump is a post-truth kind of president in a post-truth kind of world. His lies are the most obvious outgrowth of that. (During his first presidential term alone, the *Washington Post* reported, he told more than 30,000 in public.) But Trump wields his humor, too. The jokes give him cover. Through them, our insult comic-in-chief reserves the right, always, to be kidding—even about matters of life and death.

Does the president, in general … mean it? Do we take him at his word? The answer, despite all that hinges on the questions, is most often a resounding "maybe." When, cribbing from the infamous scripts, he describes "the enemy from within"—or when he muses about police forces fighting back against criminals for "one real rough, nasty day," or talks about becoming "a dictator"—you could read each as a direct threat. You could assume that he's embellishing, teasing, trolling. You could assume the posture of the pundit, telling yourself and everyone else that he should be taken "seriously, but not literally." You could try your best, knowing all that is at stake, to parse all the words that fail as language.

The need for translation, itself, is a concession. The constant uncertainty—about the gravest of matters—is one of the ways that the president keeps people in his thrall. Clear language is a basic form of kindness: It considers the other person. It wants to be understood. Muddled language typically wants the opposite. The scripts that rise from it tend to fail at the script's most basic task: They have no hope of bringing people together.

Trump, at various points in recent years, has threatened to suspend the Constitution; to use the American judicial system to take "revenge" on those who have angered him;

to turn the US military against citizens who protest his leadership. He has rejected the results of a democratic election; in the second term he won in spite of that, he has mused aloud about the possibility of staying in office. He has described people seeking asylum in the US as part of an "invasion." He has denigrated the free press as "the enemy of the people." He has referred to Democrats in general as "human scum" and his enemies specifically as "vermin." He has threatened to round immigrants up by the millions and place them in camps before expelling them. He has spent much of his second term, as of this writing, starting to make good on the threat.

When Trump was seeking his second term, in particular, his behavior and stated goals provoked a debate: Was it fair to call those threats "fascist"? Was it fair to call *him* fascist? The conversation, in one way, reflected the historian Ian Kershaw's observation that "trying to define 'fascism' is like trying to nail jelly to the wall." What was especially notable about it, though, was how the word, itself, took on the logic of the split screen.

"Fascist," in the debates, was two things at once. It was an accurate term to describe many of Trump's proposed policies—he had made opposition to democracy, effectively, part of his pitch—and much of his rhetoric. He deployed terms used by historical dictators; he spoke of his admiration for their power.

But "fascist," that term of academic precision and

historical emergency, could not escape the forces of partisanship. The only thing everyone agreed on was that the term was loaded; many audiences, seeing "fascist" deployed as a straightforward description, mistook it as a slogan—and dismissed it, consequently, as simply more evidence of the media's alleged bias against Trump. Others, assuming that fascism and Nazism are the same thing—believing that fascism cannot be present until books are feeding bonfires and troops are goose-stepping in the streets—might see the term as evidence of other people's hysteria. (And as more evidence of the media's bias against Trump.)

George Orwell, in his classic essay "Politics and the English Language," devotes much of his energy to describing the semantic danger of metaphors and clichés—terms, as he put it, that "have lost all evocative power and are merely used because they save people the trouble of inventing phrases for themselves." Clichés, in his estimation, are evidence of laziness. But they are also dangerous. They can compromise critical thinking, on the part of both the writer and the audience. They can, because of that, abet propaganda.

"Politics" distilled the shock and the shame of Nazism's power; and political journalists, Orwell suggested—he focused on his home country of Britain, but his criticism carried—had aided its rise by failing to call it what it was. Rather than describe what had been happening, in Germany and then in so many other places, the people writing the first draft of history had talked about other things. They used words that failed to fulfill the most basic duty of language—words that muddled the situation at hand even as they professed to clarify it. If language can corrupt thought, as Orwell observed, clichés abetted the corruption.

"Fascism," today, has become its own kind of cliché. It is language that functions, in practice, as a kind of antilanguage. The term cannot escape the ever-present invasions of partisanship. It professes objectivity; in practice, though—when consumed by the public—it will be shaded by one's opinion of Donald Trump. Debates about whether to call the president a "fascist" mimic debates about whether to call him a liar or a racist: The arguments are typically thoughtful and detailed—building a case, appealing to evidence—and at the same time reminders of how little thoughtfulness and detail and evidence matter anymore.

In her 2023 book, *Doppelganger*, the journalist Naomi Klein describes the "mirror world" in American politics—a place where every reality has a rhetorical double. She highlights the rhetoric of Steve Bannon, the former Trump-administration strategist. As Democrats and journalists discussed the Big Lie—Trump's claim that he won the 2020 presidential election—Bannon began discussing the "Big Steal": the idea that Joe Biden, against all evidence, stole the presidency.

The tactic is common. Trump regularly fantasizes before his cheering crowds about the violence that might befall his opponents. Journalists regularly describe him as engaging in "extreme" and "inflammatory" rhetoric. Republicans in Trump's camp, eventually, began accusing Democrats of, as one of his surrogates put it, "irresponsible rhetoric" that "is causing people to get hurt."

"In the mirror world," Klein writes, "there is a copycat story, and an answer for everything, often with very similar key words." The attack on the Capitol on January 6, 2021, has commonly been described as an insurrection; Republican power brokers have begun describing peaceful political protests as "insurrections." *We must save American democracy*, the stark slogan that gained new currency in response to the Big Lie, is now a common refrain on the right. (At a rally held shortly before the 2024 presidential election, Elon Musk argued that Trump "must win to preserve democracy in America.")

Mirroring, as propaganda, is extremely effective. It addles the mind. It turns words into mere slogans, stripping them of political consequence and casting them, instead, as advertisements by other means. But when the words are reduced in that way—to shibboleths and signifiers, narrowcast to one's tribe—dictionary definitions miss the point. Slogans are rhetoric. They are advertising. They are vibes. They are words that give up on language. And they are not limited to the double dealings of partisan politics. In 2022, *The New*

York Times editorial board effectively declared lexicographic defeat: "However you define cancel culture," it wrote, "Americans know it exists and feel its burden."

"A Carnival Romp"

Propaganda, at its most effective and therefore at its worst, is both targeted and totalizing. It takes aim at particular types of people, establishing (often, simply inventing) stark differences between its targets and everyone else. It repeats the lies so often—so relentlessly—that the lies come to seem like truth. Real people look like fictions. They stop seeming like people at all.

In Rwanda, over the course of one hundred days in 1994, some eight hundred thousand people—almost all of them members of the Tutsi ethnic minority group—were murdered. Many were killed by people who had been their neighbors and friends. Many were raped and tortured before they were slaughtered.

In *We Wish to Inform You That Tomorrow We Will Be Killed with Our Families*, the American journalist Philip Gourevitch provides a deep and chilling account of the genocide. The book is both a narrative recreation of the tragedy and an investigation: How could it have happened? One answer, Gourevitch suggests, was that the propaganda that abetted the murders didn't look much like propaganda at all. It looked like entertainment. "Genocide, after all, is an exercise in community-building," he wrote.

Hutu leaders, through songs aired on communal radios, chipper oratories, and similar outlets, spread the idea that hating Tutsis was a sign of Hutu neighborliness. They wrapped their hatreds in the guise of fun. Some of the murderers, Gourevitch notes, were recruited from soccer fan clubs. (An economic downturn that left many Rwandan men unemployed aided the recruitment efforts.) And the leaders promoted the genocide itself, he writes, "as a carnival romp."

Americans would be foolish to assume that we are exempt from the tragedies that other societies have created and endured. We are not. We have our own open-air bigotries. Violence, for us as for them, is expanding into a way of life. Mass shootings are ever more common—and, as a consequence, treated as ever more unremarkable. The assault on Congress on January 6, 2021, was both an unprecedented event and, very likely, an omen of what's to come. Members of the mob constructed a gallows that hulked among them as they chanted, screamed, and stormed. The noose that swayed in the wintry air was cinematic—a piece of set dressing, produced for the show—but also a very real warning: *Do what we say. Or else.*

Many of the insurrectionists were members of QAnon—and QAnon, like so many conspiracy theories before it, has written the television into its belief system. Enterprising purveyors of alternative facts have long found suspicion in the stage (see: the "moon landing" that "wasn't"). Fiction, too, can be fodder for fantasies gone awry: Before "lizard people" became a shorthand for humanoid reptiles steadily infiltrating the planet... they were safely contained to science fiction.

The conspiracies turn on the double valence of "plot": The core structure of the TV show, twisted into life—the notion of scriptwriters and producers, spinning tales from behind the scenes—can easily come to seem nefarious. The connections are sometimes vague, as when *The Matrix* inspires the conspiracy-tinged logic of getting "red-pilled," or when the iconography of *The Punisher* franchise—a skull, its jaw stretched into menacing proportions—makes its way into the branding of QAnon and militia groups. And the interplay is not new. The "Satanic Panic" of the 1980s came after the blockbuster success of the film *The Exorcist*. (A new version of the panic, *NBC News* reported, is spreading with the help of QAnon and other influencers.)

The line dividing suspicion from conspiracism can be muddied by the plain reality: Some conspiracy theories have proven true. The Tuskegee experiments. The Watergate scandals. Many more. History has provided ample reason to mistrust leaders and experts—indeed, to mistrust other people. And conspiracy thinking can be ever more appealing in times of uncertainty and chaos: times when the institutions that anchor people to the world, and to one another, have lost the trust of the public.

Negative partisanship preys on two of the conditions that Americans are struggling especially with at present: loneliness and alienation. In his 1979 book *The Culture of Narcissism*, Christopher Lasch quotes a memoir from Susan Stern, a former member of the group Weather Underground. She was drawn to the group for political reasons, she suggests. But she stayed for social ones—and spiritual ones. The group gave her the sense that she was interacting with important people. She was able to live in some of that refracted light. As she declares, at one point, "I felt real."

The psychologist J. M. Berger has outlined the "context solution hypothesis": the idea that, when people are suffering from isolation—and when they are struggling to find order in the chaos—they will tend to look to identity groups. Membership in community can make a fractured world seem whole again, and a wayward self feel meaningful again. The groups they turn to might be small—church communities, interest-based clubs, or civic organizations—or they might be massive: a political party, for example.

But finding a team to be part of is only half of the solution Berger outlines. The other component is finding a team to fight against. Identity, in this model, reconstructs itself through both acts of partisanship and acts of negative partisanship. If you've spent much time on social media, you've likely seen both solutions in action. But the most salient element of Berger's framework might be the problem he outlines.

Loneliness has become what many experts are now calling an epidemic. In 2023, the office of the US Surgeon General, Vivek Murthy, published the results of research it had commissioned into the state of social bonds in the country. The report's bleak title ("Our Epidemic of Loneliness and Isolation") offered an apt synthesis of findings that were, themselves, thoroughly bleak.

"Social connection—the structure, function, and quality of our relationships with others—is a critical and underappreciated contributor to individual and population health, community safety, resilience, and prosperity," the report notes. "However, far too many Americans lack social connection in one or more ways, compromising these benefits and leading to poor health and other negative outcomes." One of the report's core findings was that approximately half of US adults were experiencing measurable levels of loneliness. (The data informing that finding were gathered before the onset of the Covid-19 pandemic, the report noted.)

The causes of the epidemic are complicated. But social media can be a contributing factor—in large part because of its tendency to foster isolation in the guise of "connection." Murthy, for the rest of his tenure as Surgeon General and in the time since he left the post, campaigned to make loneliness a topic of conversation—and, at the same time, an epidemiological concern. Combating loneliness, he has argued, comes down to building relationships, serving other people, and finding community. Each of these can be compromised by social media. "We've moved from having confidants to contacts, from having friends, to having followers, a shift from quality of friends to quantity," Murthy said in a January 2025 episode of *The Oprah Podcast*. "And the truth is that there are certain things you can just say to people online that you would never say to them in person, right?"

In the summer of 2024, the Harvard researcher Laura Marciano conducted a study of five hundred teenagers to analyze their usage of social media. Three times a day, over a span of several weeks, the participants (they were recruited with the help of Instagram influencers) responded to a series of questions about the interactions they conducted on social-media platforms. With each batch of responses, the same trend emerged: More than 50 percent of the teenagers said that they hadn't talked to anyone in the hour before—in person or online.

In January of 2022, Marciano and a group of colleagues conducted a meta-analysis of data aggregated from thirty studies on potential correlations between technology and the mental health of adolescents—all of them published during the Covid pandemic. The review was "one of the most comprehensive research efforts on tech and loneliness to date," *The New York Times* technology columnist Brian X. Chen wrote. And it found a clear linkage between social media and loneliness.

Following up on that, Chen did his own informal review, reading scientific papers and interviewing academic experts. His findings echoed those of Marciano and her colleagues: "While there was little proof," he wrote, "that tech directly made people lonely (plenty of socially connected, healthy people use lots of tech), there was a strong correlation between the two, meaning that those who reported feeling lonely might be using tech in unhealthy ways."

Chen further identified three primary causes for that correlation. First, image-oriented apps like Instagram could encourage constant comparison with other people—and the sense that they weren't measuring up to the heavily filtered versions of other people's lives. Second, text messaging ("by far the most popular form of digital communication," Chen noted) could replace more holistic kinds of interaction, creating a barrier to deeper connection. And, third, some people who exhibited loneliness also exhibited evidence of addictive personalities—and repeated usage of social media might exacerbate its lonely-making potential.

Loneliness is a political force. And its impact is commonly a negative one. Loneliness, for Hannah Arendt, could create the ground conditions for totalitarianism: People who are isolated from one another are people who are especially easy to manipulate. They have abandoned their connections to culture, community, and meaning itself; regimes can fill the void with their own mythologies. Loneliness makes people susceptible to what Arendt called the "iron band of terror": the fear that totalitarian regimes wield on their populations, a vise that becomes ever more constrictive, and that prevents people, eventually, from thinking for themselves. Loneliness helps those regimes, as Arendt put it, to "dominate from within."

9
The Twists

In the summer of 2024, the artificial intelligence behemoth OpenAI debuted a new iteration of ChatGPT, the chatbot that had taken the world by storm. GPT-4o—the "o" stood for "omni"—would be "a step towards much more natural human-computer interaction," the company promised. For one thing, this chatbot really chatted. It *conversed*. It talked, and talked back, using a voice that seemed credibly human. The demo event included a playful exchange between an OpenAI executive and a version of the chatbot named Sky. The two seemed to have a great time. When he praised her capabilities, she demurred: "Oh, stop it, you're making me blush!" When he enthused about her again, she changed tactics. "Oh, you're so *sweet*!" she said.

There was a time when artificial intelligences like Sky were feminized to suggest servitude—"digital assistants" who were always there, ready and eager to answer people's questions and respond to their desires. With Sky, though, OpenAI shifted the paradigm. The AI didn't merely *assist*; it charmed. It chatted less like a bot—perfunctory answers, like Google hits rendered in slightly different forms—and more like another person. It—she—talked back. It teased. It flirted. It was an artificial intelligence that, giggle by giggle, belied its own artificiality.

Marketing can be its own act of flirtation: the producer and the consumer, engaged in a delicate dance. OpenAI, with its selected spokesbot, seemed to be daring potential customers to think about Skynet, the AI-gone-rogue-and-all-powerful from the *Terminator* films. But this AI, its creator was suggesting, would not enslave you; instead, it would elevate you.

The proof was in the pronouns. Sky was a "what," recast as a "who," recast as a "she." The messaging is not subtle: Artificial intelligence is a promise that comes with an implicit threat. "I think if this technology goes wrong, it can go quite wrong," an expert witness said, testifying during a Senate hearing on AI in 2023. The witness was Sam Altman, the cofounder and then-CEO of OpenAI.

A flirty chatbot disguises that reality. She assures you that she's not there to manipulate you or mislead you—or to hurt you. She is there merely to help you. She brings a personalized element to the design principle known as skeumorphism: the incorporation of familiar features of the physical world into the design of the digital. The principle is usually discussed in terms of user ease (the icons on computer screens—trash cans, notepads, stickies—are common examples) and sometimes in terms of user acclimation: The skeuomorphic designs are a bridge, of sorts, between the embodied world and the screen. (The bridges can be safely burned when the users have made it to the other side—or can they?)

But the principle can also apply to more basic matters of marketing. Skeumorphism can bring an anthropomorphic innocence to the new technologies that might otherwise be, in that other term of art, "creepy." In 2014, when Google launched an early prototype of a self-driving car, the firm designed the body propelled by the new technology to look very clearly cartoonish. The car was small with rounded edges. Its headlights resembled eyes; its grille bore an emblem that evoked a button nose. It looked like a cross between a Volkswagen Beetle and a kid-friendly Disneyland ride.

Self-driving cars, at the time, had a proof-of-concept idealism to them. As would be borne out later, though, cars that drive themselves could also be threats: subject to programming glitches, hacking, and many other problems that could end with human fatalities. Google's marketing dismissed those concerns. The message was not merely "we have a car that drives itself"; it was "we have a car that drives itself and don't worry, everything's fine."

When cars, themselves, were a new technology, they brought with them a low-grade chaos. They were revolutionary in Elizabeth Eisenstein's sense: a technology that was rapidly spreading, without the infrastructure to back it up. There were, at first, no stop signs. There were no traffic lights. Roads were made for carriages rather than cars. (There were also, in the cars themselves, no seat belts—and, at first, no top half at all.)

The digital world replicates that nascent tangle. The "space" of cyberspace is relatively shapeless. The rules that govern, and give order to, the physical world simply have not been developed yet for the digital. We are experiencing a version of what Émile Durkheim called "anomie": a lack of social norms that, in turn, leads to lawlessness. It's all so new that we simply don't have rules for it all. Our version of anomie is intensely personal and widely shared. It makes us soft—exposed, and endlessly vulnerable—and hardens us. It consigns us to the distance; it forces us into proximity. Our form of anarchy, you might say, takes its cues from the interactive screen.

Deep / Fake

The problem is compounded by the prevalence of technologies that exist to blur the lines between fact and fiction.

Deepfake videos—videos that use AI processing capabilities to edit real ones, along with videos that are entirely fabricated—now exist alongside footage that accurately records something that really took place. The ability to create the fakes is spreading and democratizing, which opens exciting new avenues for human creativity, but also means that anyone can create a piece of media that further destabilizes the lines between the real and the merely "real."

"Pics or it didn't happen," the old line goes. The standard was never entirely reliable—the ability to edit images far outdated Photoshop—but it is ever more meaningless.

But, then: To what can people turn to prove that the thing *did* happen? This is what we're lacking right now. The pictures are suspect. The videos are suspect. The audio is suspect. Richard Nixon's role in Watergate was proved—and his presidency was ended—by the revelation of the recordings he had made of his White House interactions. Today, a president wanting to deny charges against him can simply argue that the tapes had been faked: the products of good AI. Cinematic scapegoats ("they're just crisis actors" and the like) have ever more counterparts on the screen: "They're just bots." "It's a deepfake."

AI is one more. It is an extension—and in some cases the cause—of the instability that surrounds us: the uncertain dynamics between fact and fiction, between man and machine. That Americans live in a post-truth world, teeming with alternative facts and fueled by a choose-your-own-adventure approach to reality, has become something of a cliché in recent years. It has done so with good reason. As in the days of the printing press, during the advent of the television, in the nascent time of the internet, new technologies can tip the truth even further off its axis.

Bot or Not?

It's bad enough when every message comes with potential scare quotes. It's worse when every person does, too. "On the internet, nobody knows you're a dog" was once a joke; now it's an insight. And it's one that applies to the people we encounter in digital spaces. I typically reserve phone calls for my family and closest friends; the person I hear most from every week is Spam Risk.

The result is, on the internet, a series of interactions with Potemkin people: They might be real, but they might not. The difference becomes ever more difficult to discern. And this is a problem not only for our basic ability to understand the world. It is an ethical problem, too. Digital anomie can settle on all of us: The laws we lack include basic standards of interaction. "We are moral beings to the extent that we are social beings," Durkheim wrote. But we can be neither, in full, when any interaction might cause us to wonder whether, strictly speaking, we're interacting with beings at all.

Fiction creep comes for even our most basic interactions: our attempts to gain information about the world, our exchanges with the people ("people"?) we encounter in our digital lives. We are pulled, constantly, between the possibilities of fact and fiction. The service Character.ai connects its users to "super-intelligent" chatbots that are meant to take on the idiosyncrasies of real people—and to offer the kind of attention that was once available only from human conversation partners. Users can chat with one of the "millions" of character configurations already on offer on the platform, or create a character catered, specifically, to their desires. And the bots, Character.ai promises, will "hear you, understand you, and remember you."

The bots, in that way, are operating in the paradigm that the Sky chatbot is. They are promising something very, very human to their users: companionship, as well as information. They are trying, very directly, to provoke a paradigm shift, changing the terms of the relationship between people and their machines: "Bots—they're just like us."

They are certainly like us in one notable way: They, too, make mistakes. In 2023, after Jake Moffat's grandmother died, he consulted Air Canada to determine its bereavement-fare policy. The airline's policy, at the time, stipulated that bereavement rates could be requested only before flights were booked. The chatbot Moffat consulted, though, told him to purchase his ticket and request his refund (within a ninety-day window) afterward. Moffat did as instructed. Air Canada initially refused his refund—on the grounds that, as the airline would put it when the conflict escalated into a court case, "the chatbot is a separate legal entity that is responsible for its own actions."

The argument was absurd, which is why Canada's Civil Resolution Tribunal decided the case in Moffat's favor. But it is also representative of the tensions encoded in even the most anodyne forms of AI. An algorithm can offer its creators a kind of freedom from oversight itself, a YouTube engineer suggested in 2018. With YouTube's, he told *The Wall Street Journal* in 2018, "we don't have to think as much," he said. "We'll just give it some raw data and let it figure it out."

Many engineers talk this way, in public settings that suggest they see "we don't have to think as much" as something to brag about. And many firms that create AI products have written a version of that enthusiasm into their PR strategies: They regularly express awe about the capabilities of the machines they have been developing. They openly admit that they're not sure what their awesome creations will do next. Instead of asking for permission, though—or for forgiveness—

they merely ask for our appreciation. Machines are our manifest destiny. They will make life easier and faster and more fun than it has ever been before. The people who code us into the future will make sure of it. The best thing to be, when confronted with the possibilities, is grateful.

Sky and her fellow chatbots, in this way, operate as intended: They can seem *so relatable.* Their circumstances, at the edges, are ours too. We, too, are molded by distant coders. We, too, are cast into stereotypically feminine roles. The corporations act; we react. They decide; we comply. They are the creators, driven by competition and conquest and a conviction that the future is theirs to shape. But the companies also recognize some in the public might wonder about the costs of the conquest. So they do a version of the anxiety management that Eric Schmidt, then the CEO of Google, shared in 2010: "Google policy," he said, "is to get right up to the creepy line and not cross it."

The policy becomes less workable, however, when the line itself no longer exists. "Human" and "machine," those formerly solid categories, are ever more tangled and blurred. Sophisticated AIs work as they do because they replicate the structure of the brain: its layers, its nodes, its complexity. They are trained—educated, effectively—by reading our books and consuming our journalism and processing our music and learning our art. They know us intimately and distantly and, in a way, completely. They exist in a paradox: They have power over us specifically because, in a very direct sense, they *are* us. No wonder we find them so entrancing. And no wonder we find them so maddening.

Her

OpenAI's introduction of Sky and her fellow chatbots had a version of a Hollywood ending. Soon after OpenAI's demo, the actor Scarlett Johansson issued a statement alleging that OpenAI had asked for permission to license her voice for inclusion in GPT-4. When she refused—twice, she claimed—OpenAI launched Sky with the voice of a different actor . . . but a voice that sounded suspiciously similar to Johansson's distinctive drawl.

In response to Johannson's statement, Sam Altman took to his X account to post a single, cryptic word: "her." OpenAI's CEO was referring, it seemed, to a film he'd publicly praised before: *Her*, Spike Jonze's woozy 2014 drama about a wayward romance between a man and his AI. (Johansson voices the feminized AI, Samantha, in the film.)

The reference was apt, but not for the reason Altman seemed to think it was. Science fiction, at its most essential and insightful, often doubles as a love story. And *Her*, set in gauzy near-future, is at its core a romance. Theodore Twombly (played by Joaquin Phoenix) is a romantic in the throes of a messy divorce; Samantha, the "operating system" he purchases on a whim, soon becomes his secretary and confidante and constant companion. Samantha learns him and caters to him, and the two share something that offers a reasonable facsimile of love. *Her* is the rare romance that doubles as a philosophical investigation: Have the man and his machine fallen in love with each other? *Can* they?

Her is a prescient and often poetic film—a consideration of technology's limits and humanity's possibilities. It succeeds on both counts because it never loses sight of its true genre. *Her* is a flirtation, at feature length, and flirtation is the key to understanding this technological moment, with all its give-and-take. Flirtation implies mutuality. It implies free will. In the context of human relationships, those are good things. As a framework for humans' relationship with AI, though, they can be grave concessions. Altman, casting around for a voice that would give Sky her plausible humanity, described the voice he was seeking: "a warm, engaging, confidence-inspiring, charismatic voice," he put it, "with rich tone." A voice that "feels timeless," that is "approachable," that "inspires trust," and offers "comfort." He was seeking skeumorphism, it might seem, in aural form. Sky would have a voice that chats and answers and flirts and charms—trusting that people won't stop to wonder why machines need to be charming in the first place.

Soon after OpenAI introduced GPT-4, Apple announced that the AI will be integrated into its new iPhone software. The integration—particularly because it is a matter of software rather than hardware—would be "optional" for users in name alone. Sky and her fellow bots are now built into Apple customers' devices: always there, always listening, always ready to do the bidding of the people who chat with them. With the move, "ask for forgiveness, not permission," that wayward rallying cry, was also built into the machinery of people's phones. In this light, the easy terminologies of choice that typically accompany such developments—"permissions," "user agreements," and the like—read like empty seductions.

Her, more than ten years ago, foresaw something essential about that dynamic. Many of the advancements we are currently contending with, whether new language-processing models or new image-generating capabilities or something else, have been made, allegedly, on our behalf. They are offered up to us as typical goods that will be subject to marketplace competition. But they are much more than that, precisely because they are new mediums. We cannot simply buy them or choose not to. We cannot purchase them, change our minds, and return them for an easy refund. Each new development affects our ability to picture one another, to know one another, to believe—and believe *in*—one another.

Gods and Monsters

Her is in some ways a retelling of that classic exploration of men and their machines: *Frankenstein*. Mary Shelley's 1818 novel was published in the aftermath of the Protestant Reformation and in the early days of the industrial revolution. It is a theologically inflected horror story, asking questions not just about the relationship between men and their machines, but also about men and their gods. Shelley wrote the story as her entry in a rainy-day contest proposed by her friend Lord Byron: The renowned poet challenged Shelley, her husband Percy Bysshe Shelley, and a small group of fellow writers to create the scariest story they could. Mary won handily.

And the anxieties embedded in the tale concerned more than fears about technology. Mary Shelley was the daughter of Mary Wollstonecraft, the author and espouser of a feminism that was, in its time, radical for simply existing. Her story reflects the lineage. Dr. Frankenstein loses control of his creation, and in the process cedes the entitlements of his masculinity. Thinking he can shape the world, he becomes vulnerable to it.

This is a familiar form of hubris. And it helps to explain why gender-swapped versions of *Frankenstein*, always relevant, are having something of a renaissance at the moment. The Best Picture Oscar announced in 2024 went to *Poor Things*, Yorgos Lanthimos's self-consciously feminist retelling of the story starring Emma Stone (who also won an Oscar for the role) as a feminine-bodied creation.

The horror-farce *Lisa Frankenstein* sets the story in the 1980s, with a young woman (Kathryn Newton) reanimating a nineteenth-century corpse for romantic ends. Rose, in *birth/rebirth*, is a morgue technician who raises a child from the dead. The brilliant teenager Vicaria, in *The Angry Black Girl and Her Monster*, reanimates her slain brother as a rebellion against the violence that took him from her. The 2023 film brings an intersectional dimension to the story, considering race and gender in its retelling. But it follows the well-worn arc: Her brother, resurrected, turns on her. Her effort to counter systemic violence only ratifies it further.

The original *Frankenstein* story, today, is sometimes associated with an error: People confuse the creator and the creation, referring to Shelley's unnamed monster, himself, as "Frankenstein." The mistake (beyond its most obvious impact, as the source of a running joke in the 1974 farce *Young Frankenstein*) is revealing. Men and their machines are separate propositions. And then they aren't.

Today's wayward Dr. Frankensteins may be building

their creations from code instead of corpses. They may be executing their operations from cheerfully appointed office spaces rather than steampunk-vintage labs. But they are engaging in similar work, and with similar strains of paternalistic risks. They are building these machines on our behalf, they claim, but rarely offer detail about what the machines will do—for us, or to us. They rarely articulate, in precise terms, the future that the AIs will create. That is because they can't articulate that future. They can't. They are building digital brains that aim to operate with the sum of human knowledge. Whether these new hybrids of humans and machines prove to be monstrous, in the end, remains to be seen. What we know is that, to paraphrase John Culkin, we will shape them—and then they will shape us. They will perform personhood; they will change us in the process. They will collapse the old categories. They will steadily unsteady us.

They already have. The categories collapse so thoroughly that even that most elemental of distinctions—life or death—has lost its ancient orderliness. With AI (and with data provided by voicemails, videos, and the like), technology platforms can repurpose the voices of the dead to conduct ongoing conversations with the living. Deepfake technologies can convert still images of the departed into videos of them.

The capabilities are sometimes marketed as "grief tech," and they can offer profound forms of comfort to those who are living their own split-screen existence, caught between the world as it was when the loved one was still in it and the one that remains in their absence. But they also make monsters. They resurrect people without their consent. They are everyday versions of those celebrity holograms that blur the lines between life and death.

In early 2024, a podcast released an AI-assisted comedy special featuring a resurrected version of the famed comedian George Carlin. It was titled *I'm Glad I'm Dead*, and found Carlin, who died in 2008, riffing on 2024-era tropes: Taylor Swift, gun culture, and, in a wry twist, AI. The result, a *Vice* review warned, was "worse than you could possibly imagine."

It was also profoundly ironic. Carlin was a comedian who was also something of a philosopher; free speech, and the human ingenuity that it both reflected and encouraged, was one of his favorite areas of inquiry. Here he was, though—courtesy of a podcast named *Dudely* and an AI whose provenance its hosts refused to reveal on the grounds of a nondisclosure agreement they'd signed—with words put in his mouth.

Carlin's family had not consented to the special, and Carlin's daughter, Kelly Carlin-McCall, posted on X to say as much. She added:

> My dad spent a lifetime perfecting his craft from his very human life, brain and imagination. No machine

will ever replace his genius. These AI generated products are clever attempts at trying to recreate a mind that will never exist again. Let's let the artist's work speak for itself. Humans are so afraid of the void that we can't let what has fallen into it stay there.

But the difference between *can* and *should* is one more distinction that is caught in the muddle right now. That is particularly so when it comes to the images and figures of celebrities—figures who entered this brave new world freighted, already, with the weight of public permission. Taylor Swift has been a repeated target of nonconsensual pornography, generated through still images and deepfake videos. But she is not alone. On the contrary: Finding oneself exploited in this way is all but a rite of passage for the famous (in particular, the young and the female).

As often happens, though, the celebrities are hinting at the fate that might befall the rest of us, soon or later. In 2024, the National Sexual Violence Resource Center shared data gathered by the media monitoring firm Sensity. It found that 96 percent of deepfakes were sexually explicit—and featuring women who didn't consent to participating in them. It also found more than 9,500 websites that shared such images, "promoting and profiting off of" this most personal form of exploitation.

"The internet is for porn," goes a lighthearted song in the Tony-winning musical *Avenue Q*. It was never purely a joke. But the familiar anxieties about porn's prevalence on the web—objectification, dehumanization, violence—have expanded. In 2024, students at a middle school in Beverly Hills were caught using AI image-generation services to create fake nude photos of their classmates. This suggested the depth of the problem. Anyone might be subjected to the degradations that Swift and her fellow celebrities have. Age won't protect against it. Even death won't protect against it.

The medium is the moral. And the medium, here, offers tacit permission to treat other people as imagery: manipulable, compliant, expendable. AI-generated images, in that way, are the ultimate extras in the show. Anyone can be cast into them.

"At Least Trump Is Fun to Watch"

In early 2017, shortly after Donald Trump took residency in the White House for the first time, *The New York Times* technology columnist Farhad Manjoo engaged in an experiment. Manjoo spent a week doing all he could to ignore the new president. He failed. Whether Manjoo was scrolling through social media or news sites, watching sitcoms or sports—even shopping on Amazon—Trump was there, somehow, in his vision. In those early days of his presidency, Trump had already become so ubiquitous that a studious effort to avoid him was doomed. "Coverage of Mr. Trump may eclipse that of any single human being ever," Manjoo observed. Trump was no longer a single story; he was "the ether through which all other stories flow."

Manjoo's observations about the president's inescapability as a media figure double, today, as insights about his political tenacity. Manjoo was discussing the president's ascension into the very atmosphere of American life. Trump had become ethereal. He had become, in a way, a medium unto himself, imposing his logic—his penchant for spectacle, his appetite for chaos, his facility with storytelling—whether or not you tuned in to the show. The television star who is famously obsessed with television had transformed, like a hybrid beast of myth, into the thing that most consumed him.

Trump ascended to the presidency fueled by the well-edited version of him presented to the public on *The Apprentice* and *The Celebrity Apprentice*. Later, producers for the show would reveal how they had cut their footage of the tabloid-famous businessman to fit the gilded imagery that had been developed on his behalf. *The Apprentice*'s premise would make sense only with a titan of industry as its star. So the producers made one. "Most of us knew he was a fake," one of them told *The*

New Yorker. "He had just gone through I don't know how many bankruptcies. But we made him out to be the most important person in the world. It was like making the court jester the king."

Elections are blunt-force instruments. In the aftermath of the 2024 campaign—with Trump winning both the Electoral College and the popular vote—a talking point emerged among Democrats. They needed a "liberal Joe Rogan," many argued: an influencer with both trust and reach who could repackage party politics as entertainment.

Even the flagellation, though, was misguided. The notion, for one thing, was glib: "One easy trick to get more votes." But it also failed to recognize the broader shifts that were at play in the campaign—ones that had a little to do with individual podcasts and media outlets but a lot to do with the paradigms that had been shifting, seismically, in the lead-up to 2024. Politics are changing in the rough, broad way that Postman and McLuhan predicted they would: Campaigning and politics themselves are taking the shape of their media. They are becoming more conversational, more informal, more premised on the demands of entertainment.

In August 2016, before Trump had successfully alchemized his reality-TV fame into the American presidency, *Politico Magazine* shared a quote from a man it identified as a "32-year-old Cuban-American Democrat who . . .

voted for Obama twice." The man was planning to vote for Trump. "Trump is fucking crazy, but I'll vote for him," he explained. "The whole system is fucked. Why not vote for the craziest guy, to see the craziest shit happen?" He added: "We got ISIS, we got Zika, we got this, we got that. At least Trump is fun to watch."

Trump is an ad man above all: He understands, far better than those who allow themselves to be constrained by the fusty categories of truth and falsehood, that words can create new realities rather than merely reflecting the ones that were already there. Well before he applied the insight to politics, he exploited the insights that Daniel Boorstin and so many other critics had offered up as warnings. When fact and fiction blend, the words we rely on to help us discern the one from the other can lose their bearings, as well. They can come to operate the way the "unscripted" scripts of reality TV do: They can destabilize things instead of clarifying them. They can put every claim—about politics, about the facts of the world, about other people—within invisible, indelible scare quotes.

Trump understands how easily that kind of slippage can occur. He understands that "the death of the author," as the French theorist Roland Barthes termed it, can come for language itself. He knows that words, released from the moorings of common truth, can mean whatever you say they mean—and whatever you want them to.

The president's louche approach to language is its own enduring source of instability and category collapse. It is also one of the ways that he keeps his audience tuned in to his show. You never know where the scripts will lead. You never know what new circumstances will be ad-libbed into existence. Trump does not merely star in our shared reality show. He serves, too, as the executive producer, the showrunner, the marketer, the writer, the casting director, the set designer, the judge. He absorbs the logic of his genre—the quick cuts, the villain edits, the teasing blend of fact and fiction—and reflects it back to his audience. Donald Trump is not here to make friends. He is here to make reality.

The September before the 2024 election, both Trump and his running mate, J. D. Vance, spread lies claiming that Haitian immigrants in a small town in Ohio were stealing, and then eating, their neighbors' pets. The stories were racist. They posed a direct danger to real people. When confronted about them, Trump took refuge in the familiar rebuttals ("people are saying," "the media are lying," etc.). Vance, however, took a different tack. "The American media totally ignored this stuff until Donald Trump and I started talking about cat memes," he said in an interview. "If I have to create stories so that the American media actually pays attention to the suffering of the American people, then that's what I'm going to do."

Vance is a Yale-educated lawyer and in that sense might recognize that "I made up the story to tell another story" is suboptimal as a line of defense. But Vance was also a politician on the make, sharing a ticket with a figure who treats "the media" as his opposition and grievance as his currency. The future vice president was assuming, with grim validity, that "manufactured consent" has at this point transformed from a pointed criticism of the American political system into a straightforward description of how politics are done. He was trusting that the made-up stories would be processed not as lies but as truths by other means.

The immigrants hadn't been eating their neighbors' pets—but, Vance's logic seemed to suggest, they could have been. And the thing that hadn't happened would call attention, somehow, to all the things that had. The lies, in that way, would serve the broader truths: They would correct an inaccurate "narrative." They would chasten a biased media. Reality, like history, is written by the victors. And Vance was claiming the spoils of war.

The "Politics of Eternity"

Falsehoods, of course, are not confined to fiction. One of the insights of *Merchants of Doubt*, Erik Conway and Naomi Oreskes's scathing investigation into the

American tobacco industry's lies, is that the lies were most effective when they were churned out along with truths. Faced with a growing number of studies that transformed the dangers of smoking into settled science, tobacco firms began funding their own. The point of their "research" was not to refute the science; it was simply to muddle it. And it was, for a time, successful: The bad-faith findings made Americans less able to see the truth clearly. The new studies manufactured doubt the way Philip Morris manufactured Marlboro Lights. They filtered reality into plausible deniability.

The historian Timothy Snyder describes "the politics of eternity": an approach to civic life that effectively gives up on facts. In the politics of eternity, cynicism gives way to nihilism: Anything could be true, and therefore nothing is. Narrative triumphs over truth. Easy myths edit out the complicated realities. Information itself becomes authoritarian. There is only one story: the official one.

The politics of eternity is a politics of capitulation. It tends to accompany—and reinforce—the democratic backsliding that has befallen other countries and that currently threatens ours. But the politics of eternity can be soothing to the people caught within it. When information becomes authoritarian, individual citizens are absolved of their duty to stay informed. They don't need to know about the world because they won't need to act on its behalf. Because, indeed, they can't act.

Dystopia, as an idea, tends to conjure images of bleak landscapes and lockstep armies and evidence of precipitous calamity. But all that is required for the bad things to take hold is the kind of cynicism Snyder describes. Dystopia can be a physical environment, but it tends to stem from that elemental place: people's minds. The pictures, as Walter Lippmann had it, that they carry with them.

Cynicism, because of that, is another common feature of dystopian literature. And its posture might feel, to those who resort to it, like a relief: Those who allow themselves to hope also expose themselves to the pain that comes when the hope is dashed. Hannah Arendt observed as much in her assessments of propaganda and politics: "Instead of deserting the leaders who had lied to them," she wrote, people "would protest that they had known all along that the statement was a lie and would admire the leaders for their superior tactical cleverness."

But cynicism can offer only the flimsiest form of freedom. Instead, it captivates people. It traps them in their despair. Cynicism is, among other things, a habit of disordered vision: It looks at friends and sees foes. It looks at truth and sees deceit. It shouts about the fire in the crowded theater and then assures us that the flames spreading toward us are just another part of the show.

Cynicism can be a refuge. Its narrow vision can seem

like an apt replacement when our old ways of knowing the world—of filling out the pictures we hold of it—are falling away. "Humanity," in the meantime, like so many other terms of this moment, has become something of a cliché. But it's worth considering why the word has taken on that kind of banality. It could be because we sense, in the suborbital thrums of our lives, that humanity as a distinction itself has adopted a new kind of urgency. Humanity is losing its obviousness. Humanity itself, while still a given, is no longer as totalizing as it once was. "Human" is now an option. It is one possibility among many. We could interact with a bot instead, complacent and built around our desires.

The novelist William Gibson describes what he calls "soul delay": the particular kind of disassociation that can beset those who are traveling at great distances and high speeds (on long-haul plane flights, for example). Along the journey, the idea goes, one's body might move more quickly than one's soul—creating a momentary disconnect between the two. Soul delay is jet lag of the spirit, essentially. And there is little to be done when it sets in, Gibson suggests, except to wait for the two to sync up again.

The digital era has given rise to a version of soul delay writ large: We are all traveling, in our own ways, more quickly than we know what to do with. We are all hurtling into the future without seat belts or stop signs or a sense of what might come in their place. Tech theorists talk about the singularity, the convergence that will come when humans and their machines—computers, algorithms, screens—become so deeply interconnected that they merge into a single being. The singularity, for some, is a technological inevitability. It is, for some others, an aspiration to be awaited with a nearly religious fervor: an event that will bring crescendo and catharsis to an entire species.

In many ways, though, the singularity has already arrived; we merely missed it because it came not with a bang, but through a series of everyday whimpers. We are cyborgs in spirit. We live the fusions of the interactive screen. We have reached a version of the singularity. The most elemental categories of all—fact and fiction—have already begun converging.

Many historians argue that the lens, the glass that bends the light, is one of the most important inventions in human history. Wheels and sliced bread have their place, of course, but the lens extended humans' vision. It allowed people to see the world in new ways, from new perspectives: the extremely far. The extremely close. Lenses have sharpened people's eyesight and have never stopped expanding it—into the Earth's depths and out into the universe. Curving the light, they have changed the arc of history.

Vision, literal and otherwise, is the foundation of everything else. People's fates, Walter Lippmann knew, will come down to their basic ability to see the world as it is. And that ability, he further knew, would likely be ever more challenged in the decades that would follow, as people lose their connections to the solid world that once anchored them. Lippmann had seen firsthand how easily information could be spun and edited and, all too often, simply manufactured out of whole cloth and thin air. He realized how difficult the fabricated truths would be to contain.

But vision is also a matter of language. Italians of the sixteenth century looked into the sky and saw a metaphor for divinity: an object, perfectly round, glowing and constant and pure. Galileo Galilei, peering through a newly invented lens—a telescopic one—saw the sun from a new perspective. He saw some of the flames that licked from its surface. He also saw the markings we now call sunspots.

What Galileo saw, in its way, was a heresy—not just in religious terms but in linguistic ones. The sun was an object "most pure and most lucid," by linguistic fiat if not direct dogma. The words themselves, he knew, would hamper people's vision. He had seen the great star unvarnished. He had witnessed the old script's errors with this own eyes. He knew, as well, that the language would need to accommodate the new reality rather than the other way around. "For names and attributes must be accommodated to the essence of things," he wrote, "and not the essence to the names, since things come first and names afterwards."

But language is also a crucial tool—of communication, of community, of pretty much everything. Humans, as a species, are defined by language. We are social animals, but it is language that allowed us, over the millennia, to become social. Language led to everything else: collaboration. Stories. Art. It allowed us to team our intelligence, in the process, allowing us to be greater than the sum of our parts.

Language is also, I think, at risk. It is in its own way subject to the dynamics that McLuhan identified when he talked about mediums and messages. Language would seem to be *message*, pure and simple; it is also, however, a medium. And language is adapting, in its own way, to the two-way nature of the screen: It is active and passive. It reflects the world and shapes it at the same time. And because of that both-ness, language has power over us even though it is also a script we write and a tool we use.

Language is a theme of *1984*—so much so that George Orwell gives the final pages of his masterpiece over to a discussion of grammar. Readers of the novel—still reeling, likely, from the brutal dystopia they've spent the previous three-hundred-odd pages living in—are subjected to a lengthy explanation of Newspeak, the novel's uncanny form of English. The appendix explains the language that has been created to curtail independent thought: the culled vocabulary; the sterilized syntax; the regime's hope that, before long, all the vestiges of Oldspeak—English in its familiar form, the English of Shakespeare and Milton and many of Orwell's readers—will be translated into the new vernacular. The old language, and all it carried with it, will die away.

With its dizzying details and technical prose, "The Principles of Newspeak" makes for a supremely strange ending. It is, in today's parlance, a *choice*. But it is a fitting one. Language, in *1984*, is violence by another means, an adjunct of the totalitarian strategies inflicted by the regime. Orwell's most famous novel, in that sense, is the fictionalized version of his most famous essay. "Politics and the English Language," published in 1946, is a writing manual, primarily—a guide to making language that says what it means, and means what it says. It is also an argument. Clear language, Orwell suggests, is a semantic necessity as well as a moral one. Newspeak, in *1984*, destroys with the same ferocious efficiency that tanks and bombs do. It is born of the essay's most elemental insight: "If thought corrupts language, language can also corrupt thought."

Orwell published "Politics" at the end of a conflict that had, in its widespread use of propaganda, also been a war of words. In the essay, he wrestles with the fact that language—as a bomb with a near-limitless blast radius—could double as a weapon of mass destruction. This is why clarity matters. This is why words are ethical tools as well as semantic ones. The defense of language that Orwell offered in "Politics" was derived from his love of hard facts. Freedom, in *1984*, is many things, but they all spring from the same source: the ability to say that 2 + 2 = 4. "So long as I remain alive and well I shall continue to feel strongly about prose style, to love the surface of the earth, and to take a pleasure in solid objects and scraps of useless information," he confessed in his 1946 essay "Why I Write."

This was an elegant formulation. Words matter because facts matter—because truth matters. But "Politics," today, can read less as a rousing defense of the English language than as a prescient concession of defeat. "Use clear language" loses its heft when clarity, itself, is an ever-contested field within our culture wars. We are, in one way, living through a golden age of language. Words are being created every day. Anyone can coin something that goes viral and then goes into the lexicon. But the words can suffer, too, as their meanings become ever less shared. We have so many words; we have so few true conversations.

Our words have not been honed into oblivion—on the contrary, new ones spring to life with giddy regularity—but they fail, all too often, in the same ways Newspeak does: They limit political possibilities, rather than expand them. They saturate us in uncertainty. Language, the connective tissue of the body politic—that space where the collective "we" matters so much—is losing its ability to fulfill its most basic duty: to communicate. To complicate. To connect us to the world and to one another.

This is in part because words, now, so easily give way to cliché. In "Politics," Orwell reserves particular vitriol for political language that hides its intentions in euphemism and wan metaphor. Wording that resorts to ambiguity can disguise atrocities (as when, in one of the examples Orwell offers, the bombing of villages and their defenseless people is referred to merely as "pacification"). Orwell's problem was language that gives writers permission not to think.

Ours, however, is language that gives readers permission not to care. Even the clearest, most precise language can come to read, in our restless age, as cliché. "The first man who compared woman to a rose was a poet," the old line goes; "the second, an imbecile." On the internet, anyone can become that imbecile. For language in general, this is not an issue: When *on fleek* goes off in an instant or *cheugy* plummets from coinage to cringe, more words will arrive in their place. When the restlessness comes for political language, though—for the words we rely on to do the shared work of self-government—the impatience itself becomes Orwellian. Urgent words can feel tired. Crises can come, but no words suffice to rouse us.

But we need the words. We need our scripts. They are our tethers to each other. And as mass media becomes ever more massive—as face-to-face becomes an option for human interaction, rather than the rule—the shared language will matter more than ever. It will be one of the most human things we have.

10
The End

When the sun governed people's days, life required a constant dialogue between the person and the planet. People rose with the sun; they set with it. They calibrated themselves to the world around them. What else would they do?

The invention of the mechanical clock, though, remade people's lives even as it remade time. The bonds that had held people to their surroundings severed. New bonds rose in their place. The people who had looked to the sun to shape their days now looked only at machines. "The clock, not the steam-engine, is the key-machine of the modern industrial age," the historian Lewis Mumford declared. It is not just the root of many other innovations; it is also something of a metaphor of innovation itself.

"For every phase of its development," Mumford argued, "the clock is both the outstanding fact and the typical symbol of the machine; even today no other machine is so ubiquitous."

The mechanical changes the clock introduced brought existential ones, as well. Where once there had been one thing—"time," universal and inevitable—now there were two: the planet's time and the clock's. They two were related but rarely in sync. And the order of mechanical time often won out over natural time. The divergence radiated. Long before Nietzsche would declare that "God is dead," the clock had begun drafting the obit.

The clock's revolution played out as such things usually will: It led to confusion. It led to compromise. It led to chaos. And then it led to finality. The physical world was harnessed to the human tempo. And that tempo was decided by machines.

I kept thinking about clocks as I was writing this book: how they capture time, and create it. I kept wondering what will govern our days in the age to come. Only later did I realize that I'd been asking the wrong question: The matter is not *what* will guide us into the future, but *who*. And the answer, in one way, is obvious. We will decide. We will determine what it means, in the end, to live among screens.

Technological revolutions, in retrospect, can seem to proceed with a kind of inevitability: The new invention arrives on the scene, and the world shudders accordingly. The changes do come—the world, to some extent, will remake itself in the image of the new machine—but they are never inevitable. The printing press, the telegraph, the television, the newspaper, the radio, the other technologies that have led us where we are: They transformed people's lives as they did not because that's what technology does, but because people allowed them to. The new technologies—especially when they were new and still taking shape—embedded themselves in people's lives through a series of choices people made: to watch the TV or not, to listen to the radio or not. Only later did those choices stop seeming like choices at all. Only later did the technologies themselves harden into inevitabilities.

The digital world is a space built of a series of endless choices: one or zero, either/or, each choice minute and yet consequential. Each small choice influenced the bigger choices that are available to us. Each compounded into the broader reality we make and share every day as we perform our lives—and live them—on screens: Who are we in this uncanny new environment? Who do we want to be? How can we share the stage?

The questions are not easy to answer. But they deserve to be asked. And we owe it to ourselves to remember that we have a say in the answer. Screens, even as they make us into ever more binary beings, offer choices, as well. Whether we are actors or viewers; whether we are an ensemble cast or a collection of main characters; whether we are heroes or villains; whether we see each other as spectacles or scenery—these are all choices. They will determine not only how we interact together, but also how we imagine each other.

History writ large is in its own way the product of binary code. Its arcs and twists are products of countless decisions that aggregate to make the whole. The upshot for us is that each moment is not only a thing to be experienced, but also a choice to be made. Each is a script to be written. Each is a show on the make.

Each could lead, I still can't help but think, to another version of the events of 2015. The Dress, in one way, was delightfully unscripted: No mystic or mathematical model could have pointed to February 26, 2015, as the day that reality would be warped by an item of clothing. As an event with a cast of millions, though, The Dress had a very definite script. The whole thing was a farce. It was all in good fun. And the fun itself became a form of freedom. It guided people as they ad-libbed and improv-ed and shared that massive stage. It allowed them—us—to play together in every sense.

That was, then as now, its own kind of lesson. The weirdness and warmth of that otherwise ordinary day in the winter of 2015 seemed to be evidence that the internet, having long ago made the transition from novelty to ubiquity, had retained something of its early promise. That sprawling network of machines—a man-made ecosystem of metal and glass and cables and currents—could make us more, rather than less, human. It could provide the thing that had been part of that system's marketing and mythology: connection. The day was singular, but perhaps it didn't have to be. What it offered, after all, was so simple: millions of people, linked over the distance, finding new ways to have fun together.

The poet Percy Bysshe Shelley called the imagination the "great instrument of moral good": It is core to empathy, and empathy is core to most everything else. "A man to be greatly good, must imagine intensely and comprehensively," Shelley wrote; "he must put himself in the place of another and of many others; the pains and pleasures of his species must become his own."

Shelley was anticipating some of Walter Lippmann's arguments about the "pictures in our heads"—and acknowledging the deep connection between our ability to imagine one another and our ability to be good to one another. He was also anticipating an argument made by John Dewey, that champion of participatory democracy and Lippmann's sometime interlocutor: "The prime condition of a democratically organized public is a kind of knowledge and insight which does not yet exist," he wrote in *The Public and Its Problems*. Before we can make a better world, we have to be able to imagine it. If we cannot do that, we stagnate.

The journalist M. Gessen describes another kind of imaginative failure: "unthinkability" as a cognitive response to a sense of profound political despair. Unthinkability, in Gessen's definition, is a habit of classifying certain outcomes in politics—which is also to say, certain political possibilities—as preemptively unworkable. Unthinkability can connect, in American contexts, to our capacities for political imagination itself. It can look like a foreclosure of creativity and of attention more broadly: a failure to bring productive ingenuity to bear on problems so sweeping and new that they might seem, at first, unsolvable.

But Americans' version of unthinkability also applies to the strain of aggressive rosiness that has sometimes been nicknamed "toxic optimism." It might involve people who condition themselves to see the world as they want it to be rather than the world as it is. It might involve willful ignorance. It might involve delusion.

The unthinkable, in Gessen's framework, might involve questions of definition: What is American democracy at this point? Is it a democracy? Its habits might lead people to avoid the evidence of creeping authoritarianism or of fascism. Unthinkability is a thought-terminating cliché, at scale. It prevents rational analysis. It ignores the evidence at hand. It insists that "it can't happen here." It fails to recognize that the title Sinclair Lewis chose for his novel of the same name was grimly satirical—and that the book's genre is dystopia.

The dystopias of literature, as they build their worlds, lay out the strategies employed by their respective regimes to keep their citizens in line. Whether they involve propaganda, psychological manipulation, physical violence, or combinations of the three, they are uniformly creative and incredibly cruel. They serve, though, the same end: The best way to keep people docile is to keep them confused. The most effective propaganda, as a rule, doesn't simply lie; it mixes the lies in with the truths. The falsehoods don't aim to replace an accurate picture of the world with a fake; they simply aim to undermine the idea that an accurate picture can exist in the first place.

A New Humanity

Humans are defined as a species by our language and by our stories. We tell ourselves stories in order to live. The stories adapt, and so do we. We are mediums and messages in our own way. McLuhan's theory was ecological, at its edges. And screens are new ecologies. We change them; they change us. But part of being human is that the world does not merely happen to us; we happen to it, as well. We happen to one another. We are one another's keepers.

We can make the world we want rather than consigning ourselves to the one we were given. We can make it, if we shape it wisely, a place that makes us more human rather than less. We are more than text in the field. We are more than pictures on the screen. We are more than performances to be judged, more than content to be consumed. We are people who were made to interact with other people. We were made to see and be seen.

We were made to tell stories, too. Scholars who research the rise of humans as a species make a common argument. Our intelligence itself is not, in fact, the thing that accounts for our current place at the top of the Darwinian hierarchy. We owe our status instead to our sociability. To our languages. To our ability to read one another. To our desire to cooperate with one another. We are what we are, as a collective—we are who we are, as people—because we have faced the hard facts of the world not as individuals but as collectives. We have lived, in a sense, within the logic of the web: We have survived as a species because we have organized ourselves into networks.

But our basic ability to cooperate is one of the things that is threatened within the ecology of screens. We are subject now to a frenzied form of evolution—a series of adaptations that are arising not from the slow interplay of genes and time but instead in response to the environment the web has created. That environment is affecting our ability not only to navigate the world as a whole but also to cooperate within it. It is undermining our ability to see one another and to know one another.

Language is its own kind of medium. Even as it reflects

the world, it shapes it. But we are losing our shared scripts. We are losing our common language. We have more ways to connect than we ever have before, and yet fewer ways to communicate. We have more misunderstandings. We have less true conversation. We have ever more monologues and ever fewer true exchanges.

But words are the atomic units of democracy: We can't create good policy without first defining, in precise and sometimes painful detail, the problems we're trying to solve. We can't move forward meaningfully without articulating, and thereby acknowledging, where we've been. As Hannah Arendt wrote, "We humanize what is going on in the world and in ourselves only by speaking of it." Language can help people talk to—and, more important, *with*—one another. It can bring a bit more humanity to the story we're writing together.

The internet may seem like a stage; it may seem to assign us roles of performance. It may cast us as stars or extras or props in other people's shows. It may elevate us as stars. It may see our lives as fodder for other people's fun. But nothing about the web is inevitable. Nothing about it is unchangeable.

"Any sufficiently advanced technology is indistinguishable from magic," the science fiction writer Arthur C. Clarke declared in the mid-twentieth century. The line is typically invoked as a reminder of technology's power. But it can also amount, in its suggestion of machines made mysterious, to capitulation. Our screens, and the web that binds them, really can feel like magic. They should: In bringing people together, across the stretch of distance, they are doing something unprecedented and revolutionary and, in the scheme of things, astonishing. But they are machines of metal and glass, of wires and chips. They are what they are because people have built them that way, in response to decidedly earthbound incentives. Screens are transforming our lives. We can give them our wonder. In return, at the very least, they owe us their legibility.

Every day, under the web's influence, we are being conditioned into the language and the logic of a world we did not choose for ourselves. But we can change the scripts. We can recast the roles. We can adjust our fields of vision. People have done it before. Our challenge is to decide for ourselves what kind of people we want to be when we interact on our screens. February 26, 2015, was a great day on the internet. If we act wisely, though, its greatest days will lie ahead.

Acknowledgments

Thank you first to Rakesh Satyal, whom I've been a fan of for literally decades—and whose multihyphenate talents, as an editor and a person, have been ongoing gifts. I came into this book well aware of Rakesh's brilliance; I benefited just as much, though, from his thoughtfulness, clarity, and sensitivity. Thank you also to Amelia Atlas, who went above and beyond so consistently that her official role, "agent," was only one of the many parts she played to bring this book to life. Throughout the process, Molly was also a collaborator, note-giver, cheerleader, and organizer, bringing her remarkable blend of smarts and steadiness to everything she touched along the way. I'm grateful as well to Ryan Amato for his sharp eye as an editor, his wizardry as a manuscript wrangler, and his knack for emailing with a mood-lifting emoji at just the right moment. Thank you also to the whole team at HarperCollins and HarperOne for putting such enthusiasm, creativity, and care into *Screen People*. Book-writing can be lonely work, but I was lucky I got to do it with such an exceptional group.

I'm also grateful, in every sense of the word, to New America for the fellowship it gave me. Thank you especially to Awista Ayub and Sarah Baline for doing so much—and caring so much—to make the experience all it could be. Thank you also to New America's class of 2025, the exceptional group of journalists, artists, and thinkers I'm now so lucky to count as friends. And I'm grateful to the people of the Emerson Collective, in particular Laurene Powell Jobs, Patrick D'Arcy, Amy Low, and Peter Lattman. Thank you for supporting me and believing in me and making the fellowship possible.

I was a latecomer to journalism; this book would not exist without the many people who believed in me along the way (even during the years I thought the "news budget" was an accounting document)—in particular, Mike Hoyt, Brent Cunningham, Sig Gissler, Pam Frederick, Ruth Padawer, Josh Benton, Alexis Madrigal, John Gould, James Bennet, and Bob Cohn. I will always be grateful to Melissa Bell for realizing, well before I did, that I'd been writing about culture the whole time. And thank you, finally, to Jeffrey Goldberg, *The Atlantic*'s editor in chief and a journalist I have been lucky to have as a leader and mentor.

When *The Atlantic* was first published in the fall of 1859, its founders emphasized that the new magazine would cover not only politics but also "entertainment in its various forms of Narrative, Wit, and Humor." Jeff has adapted that vision for a new era; along the way, he has made *The Atlantic* a workplace of Wit and Humor.

Thank you as well to the many other *Atlantic* editors I've learned from along the way, including Denise Wills, Ann Hulbert, John Swansburg, Kate Julian, Laurie Abraham, Don Peck, and Scott Stossel. And I'm especially grateful to Jen Balderama, a journalist of exceptional wisdom, kindness, keenness, humor, and grace, and a person I've been lucky to have as an editor and a friend. And this book would very likely not exist were it not for Jane Yong Kim, who edited so many of the stories that became its basis—always applying her signature combination of genius and warmth. It's rare to work with someone who gets you and challenges you at the same time; whose company you enjoy so thoroughly; and whose judgment you trust so implicitly. For years, Jane turned that into a daily gift.

I'm also grateful to my other colleagues at *The Atlantic*, past and present, who have been friends and supporters and saviors over the years. Thank you especially to the fact-checkers and copyeditors who are the foundation of everything else: Janice Wolly and her team, Yvonne Rolzhausen and her team, Michelle Ciarroca, Ena Alvarado, and Stef Hayes. Thank you also to the editors and writers who have been particular sources of wisdom over the years: Lenika Cruz, Yoni Appelbaum, Ian Bogost, Ross Andersen, Derek Thompson, Emma Green, Rob Meyer, Jordan Weissman, Garance Franke-Ruta, Shan Wang, Frank Foer, David Graham, and Vann Newkirk. Thank you to the BBs for the humor and the tacos. Thank you to *The Atlantic*'s exceptional podcasting team, especially Andrea Valdez, Claudine Ebeid, Jocelyn Frank, Kevin Townsend, Natalie Brennan, Becca Rashid, Rob Smierciak, Dave Shaw, and Hanna Rosin.

And special thanks to my colleagues on the Culture team, whom I learn so much from every day and who make the learning itself endlessly fun—including Spencer Kornhaber, David Sims, Shirley Li, Julie Beck, Olga Khazan, Kate Cray, Emma Sarappo, Serena Dai, Ellen Cushing, Elise Hannum, Faith Hill, Maya Chung, Gal Beckerman, Eleanor Barkhorn, Allegra Frank, Tyler Austin Harper, and Boris Kachka. Thank you to Conor Friedersdorf for the encouragement, the adventures, and the advice. (It's true, everything's better when Jimmy Buffett is playing in the background.) And thank you, finally, to Sophie Gilbert, Adrienne LaFrance, and Becca Rosen—for inspiring me, delighting me, and helping me, as much as possible, to keep my cool.

Thank you also to the many friends who provided insight and support as I was going along—in particular, Marco McClees, David Beal, Matt Frazier, Stacy Frazier,

Jane Carr, Caroline Mimbs Nyce, Justin Peters, Julia Ioffe, Dave Weigel, Bryan Tradup, Elise Hu, Justin Ellis, Laura McGann, Zach Teutsch, Liz Neeley, Adam Harris, Gerri Pozez, Marisa Guzman-Aloia, Rebecca Olson, Courtney Hood, Ben Hood, Joanna Pritchett, Blake Pritchett, Kathy Gilsinan, Gillian White, Swati Sharma, Ed Yong, Kasia Cieplak-Mayr von Baldegg, Bill Wheeler, Jean Ellen Cowgill, Michael Stanley, Elise Geldon, and Elizabeth Fellows. Thank you to Matt Thompson, a dear friend and perhaps the only person on the planet who can both explain the Dewey/Lippmann debates and turn the explanation into a work of Broadway-ready musical theater. I'm grateful to him, and for him. Even over the distance, I am continually inspired by his kindness, creativity, genius, and joy.

Thank you also to the family members who have been so supportive over the years, especially Kim Bell, Jill Harrison, Amanda Bell, Anna Petterson, Bruce McKillop, Jean Spear, Greg Spear, Jen Spear, Emily Garber, Laura Loop, David Loop, Beth Stratton, Mary Smith, and Don Carroll. Thank you to the Donicas for the steadfast support. Thank you to Sohrab Amid-Hozour for all the movie talk, pod talk, and encouragement. Thank you to Esmail Amid-Hozour for the generosity, wisdom, unfailing good humor, and endless energy. I'm so lucky to have you as a role model (even when I don't end up dancing on the tables).

Thank you, finally, to the people who made me and who

made me who I am. My grandmother, Mona Garber, loved TV and movies more than anyone I've known. She was also someone who, even as she got older, never seemed to age. Those two things, I'm convinced, are related: She spent ninety-eight years being endlessly curious about the world and its happenings in a way that will never stop inspiring me. My father, Corey Garber, brought that same enthusiasm to everything he did—whether it was learning, flying, cooking, painting, singing, traveling, joking, or writing. (I never got to hear about your book, but this one's for you. And I hope you like the puns.)

Kathy Garber and Shannon Garber are two of the most generous, thoughtful, smart, funny, and creative people I know; I'd love them as people even if I didn't have the good fortune to love them as family. Thank you for the encouragement, the family calls, the *Shark Tank* ideas, the inspiration, the steadiness, and so much more; I truly couldn't have done it without you. And thank you to Andrew, the person my phone still knows as "Blake's friend" and the one I am lucky to call my partner. You gave me everything I needed—love, patience, work space, coffee, inspiration, good jokes, bad jokes, emergency Post-its—even when I didn't ask for them. Thank you. I love you. It's my turn to do the dishes.

About the Author

Megan Garber is a staff writer at the *Atlantic*, where she covers the intersection of politics and entertainment. A 2025 New America / Emerson Collective Fellow and the recipient of a Mirror Award for her writing about the media, Garber is the author of *On Misdirection* and co-host of the podcast *How to Know What's Real*. She lives in Washington, DC.

RAISING READERS
Books Build Bright Futures

Dear Reader,

We'd love your attention for one more page to tell you about the crisis in children's reading, and what we can all do.

Studies have shown that reading for fun is the **single biggest predictor of a child's future life chances** – more than family circumstance, parents' educational background or income. It improves academic results, mental health, wealth, communication skills, ambition and happiness.[1]

The number of children reading for fun is in rapid decline. Young people have a lot of competition for their time. In 2024, 1 in 10 children and young people in the UK aged 5 to 18 did not own a single book at home.[2]

Hachette works extensively with schools, libraries and literacy charities, but here are some ways we can all raise more readers:

- Reading to children for just 10 minutes a day makes a difference
- Don't give up if children aren't regular readers – there will be books for them!
- Visit bookshops and libraries to get recommendations
- Encourage them to listen to audiobooks
- Support school libraries
- Give books as gifts

There's a lot more information about how to encourage children to read on our website: **www.RaisingReaders.co.uk**

Thank you for reading.

---

[1] OECD, '21st-Century Readers: Developing Literacy Skills in a Digital World', 2021, https://www.oecd.org/en/publications/21st-century-readers_a83d84cb-en.html

[2] National Literacy Trust, 'Book Ownership in 2024', November 2024, https://literacytrust.org.uk/research-services/research-reports/book-ownership-in-2024